Here's what *Magnetic Rever*...

"Another blockbuster. Sc... t
again. Like his first book, ...,
Magnetic Reversals and E ... y
every person in the U.S. and the world. Bob's new book is often scary, but its solid foundation in honest science informs us that a magnetic reversal may alter life or even extinguish it for many years to come. This is the real thing --- what the heavens have in store for you and me and Planet Earth."

—Lou Guzzo, author, with Dixy Lee Ray, of *Trashing the Planet*, & former managing editor of The *Seattle Post Intelligencer*

"Sheer scientific drama" Wow! A Terrific Book" I didn't think Robert W. Felix could top his previous book for sheer scientific drama, but he has. Like a nuclear explosion, Felix blasts previous theories of dinosaur extinctions and other aspects of life on Earth to break through to an astonishing truth."

—Alan Caruba, member National Assn. of Science Writers & weekly commentator on climate and related issues

"A truly stunning book about the gripping history of Planet Earth. Never again will the reader be able to claim ignorance in matters of where we came from and how it all started. Move over Darwin."

—Hans Schreuder, analytical chemist (ret)

"Easy to read. Provocative. I liked the book. Kept my interest all the way through. I agree with the concept."

—Ian Kellman, M.D.

"Poor Darwin, after decades of being almost universally acclaimed as the genius who explained the origin of mankind and just about everything else, he's suddenly being beat about the head by researchers unafraid of taking issue with what passes as the current wisdom. Robert Felix shows convincingly that far from evolving step by step over millions of years, when new species of life appeared, they appeared whole and entire, and they did so after their predecessors were suddenly extinguished in a series of mass extinctions. Take that, Charles Darwin."

—Philip Brennan, veteran journalist for Newmax.com and one-time Washington columnist (Cato) for *National Review*

"The ideas you put forward are important and need serious consideration. Your reasoning and research are thorough, and the preponderance of evidence strongly suggests that "something major" is going on. The sooner we truly understand it, the better off we all will be."

—Michael R. Cohen, Ph.D. - Applied Physics

"Eye Opening!" I enjoyed this book so much that I reread it this weekend. Being raised and schooled in the current dogma of Darwin and evolution, I had a problem with the title. But after reading Robert's book, I felt like he removed the Vaseline from my glasses. All the loose ends and holes in current theory evaporated and became so easy to understand that a child could comprehend his theory. This book covers complicated subjects and the truth revealed is so simple that it now seems obvious."

– John Tregidga, Designer/Inventor

MAGNETIC REVERSALS
and
EVOLUTIONARY LEAPS

MAGNETIC REVERSALS
and
EVOLUTIONARY LEAPS

THE TRUE
ORIGIN
OF SPECIES

ROBERT W. FELIX

Copyright © 2009 by Robert W. Felix
All rights reserved. No part of this publication may be reproduced or transmitted in any form or by any means, electronic or mechanical, including photocopy, recording, or any information storage and retrieval system now known or to be invented, without permission in writing from the publisher, except for review purposes in a magazine, newspaper, or broadcast.

Cover graphic design by Victoria Michael, Aftershocks Media
Cover photo: "Great Red Aurora," by Jan Curtis, Laramie, WY
Author photo by Bryce Mohan

Library of Congress Control Number: 2008933279

Publisher's Cataloging-in-Publication Data
(Provided by Quality Books Inc.)

Felix, Robert W.
    Magnetic reversals and evolutionary leaps : the true origin of species / Robert W. Felix.
    p. cm.
    Includes bibliographical references and index.
    ISBN-13: 978-0-9648746-7-1
    ISBN-10: 0-9648746-7-9

    1. Evolution (Biology) 2. Geomagnetic reversals.
I. Title.

QH371.3.G46F45 2008   576.8'2'01538727
                               QBI08-600197

Printed in Canada
C 10 9 8 7 6 5 4 3 2 1

Sugarhouse Publishing, P.O. Box 435, Bellevue, WA 98009

CONTENTS

Preface

1 **Suddenly** 19
Evolution works suddenly · *Natura non facit saltum* Nature does take leaps · Geomagnetic reversals · Geologic time scale · Squishy blobs · The Cambrian Explosion · Why haven't sharks evolved? · Insect life exploded · The Coal Age

2 **Life of a Soldier** 27
New kinds of ammonites · Twiddling their evolutionary thumbs · Abominable angiosperms · The Fern Spike No known ancestors · Gordian Knot of Geology · A half-inch-thick layer of clay · Newcomers arrived fully developed · Bigger animals savaged

3 **A Deadly Mistake** 39
Seventy-five percent of all species go extinct · Tree pollen almost disappeared · Sea dwellers almost decimated · Ocean temperatures shot upward · Black muds · Larger animals at risk · Iridium · Reversed magnetic polarity

4 **The Irony** 47
Radiation and mutation · Geomagnetic reversals · Magnetism and the Curie temperature · Magnetic excursions aborted reversals? · Strontium · Dramatic spikes in radioactivity · Beryllium-10 · A layer of black carbon

5 **Half a Wing** 55
Top predators at risk · Sluggards slipped across the boundary · Living in the fast lane · How to avoid the duckbill's fate · What good is half a wing? · Mutant monsters · Babies in the womb much more sensitive

viii   CONTENTS

6   Oops   **63**
It's not evolution, it's creation · Dinosaurs just looked different · Five horns on its ugly face · Ears emblazoned on their rears · Some dinosaurs had gizzards · We showed up on this planet fully developed · No intermediate forms A million years with no change · Geomagnetic reversals played an important role · Apes and humans branch apart

7   Ruler of the Universe   **75**
Magnetic star · Solar flares · Northern Lights · Sunspots Sunspots and magnetic reversals · Electrified particles rain to the earth · The magnetosphere protects us · Converting energy to matter · Accelerating atomic particles to the speed of light · Carbon-14 · Sunspots colder than the Sun

8   Pacemaker of Creation   **87**
Radioactivity and our galactic orbit · Radioactivity and our celestial orbit · Ice ages and equinoctial precession Precession of the equinoxes · Geomagnetic field strength is falling · When will the reversal occur? · Upcoming magnetic reversal · It could reverse tomorrow

9   Magnetic Reversal Cycle   **97**
Laschamp magnetic reversal · Gothenburg magnetic excursion · Mono Lake magnetic excursion · Lake Mungo magnetic excursion · Magnetic reversal/excursion chart Geomagnetic reversals and ice ages · Catastrophes in sync with equinoctial precession

10   Tunguska   **105**
1,000 Hiroshima bombs · Exploding meteorite? · Tektites Chicxulub Crater · Iridium at other extinctions · Iridium at the base of a layer of coal · Excess osmium · Strontium spike at the K-T extinction · Carbon forms in the sky Beryllium-10 · Atomic bombs release carbon-14

11	**Hundreds of Thousands of Tunguskas**	**119**

Underwater explosions · Explosions in the sky · Einstein's theory applied · Creating new matter · Ten thousand times more carbon than normal · Tremendous oil catastrophe Carbon from the sky · Standing trees · Radioactivity and oil · Natural nuclear explosions · Creation · Electricity ruled the world · Dinosaurs and uranium

12	**Carbon Rain**	**133**

Rivers of oil · Bitumen raining from the sky · Why not here? · Black shales laced with carbon · Carbon from the sky · Black shales are highly radioactive · Natural nuclear reactions

13	**Dinosaur Tombstones**	**143**

Miniature dinosaur tombstones · Diamonds produced in the sky · Diamonds in the sky 12,000 years ago · Nuclear irradiation · Big Eloise · Iron balls from the sky · Radioactivity 2,000 times normal · Moving 4,000 miles per hour · Buckyballs · The Carolina Bays · A magnetic excursion · THEMIS · Magnetic ropes · Giant paw prints What are diamonds made of? · Diamonds are produced with explosives · Dinosaurs buried in coal · Z coal

14	**What now?**	**161**

The proof lies all around us · Will barrels of oil rain on my head? · It's not all doom and gloom · What can we do to prepare? · Fall-out shelters? · Our ancestors made it through

Bibliography	**169**
Index	**183**

LIST OF ILLUSTRATIONS

Geologic Time Scale	22, 23
Magnetosphere	81
Precession of the equinoxes	91
Magnetic reversal/excursion chart	100
Carolina Bays	151

ACKNOWLEDGMENTS

My thanks to all of those terrific people who gave me so much help and encouragement along the way.

I'd especially like to thank Lou Guzzo for his unwavering support. I remember my first meeting with Lou, when I asked him to read the manuscript for my first book, *Not by Fire but by Ice*. He called back the very next day, and I could hear the excitement in his voice. He explained that just before she died, Dixy Lee Ray, former chairman of the Atomic Energy Commission, had confided that her next book would be about magnetic reversals because they were so important.

Lou told me that my book would one day be called the most important book of the century. I don't know whether that will prove to be true, but it sure felt good. He has been pushing and prodding me now for ten years to finish *Magnetic Reversals and Evolutionary Leaps*, and I can't begin to describe how much his belief in my writing and my theories has sustained me.

Thanks to my sister and brother-in-law, June and George Mona, for their valuable insights and editing prowess. And thanks to my wife Pattie for her continued enthusiastic support and thoughtful feedback.

Thanks to Hans Schreuder, Alan Caruba, Phil Brennan, John Tregidga, Michael Cohen and Ian Kellman, MD, for reading the manuscript and for their valuable suggestions on how to make it better. Thanks to Eric W. Strom of the USGS for helping me find the info on the Carolina Bays.

And finally, thanks to Mike McEvoy, Dan Hammer and Kenneth Lund for their support over the years.

PREFACE

During most of my first 50 years, I accepted the idea of evolution almost unquestioningly. I didn't see how any thinking man or woman could come to any other conclusion. But as I conducted the research for my first book, *Not by Fire but by Ice*, a few bits of contrary information began seeping into my consciousness.

I learned that many geomagnetic reversals—far more than could be dismissed as mere coincidence—had occurred in sync with mass extinctions. And many of those magnetic reversals had occurred in sync with our planet's descent into catastrophic glaciation.

I learned that paleontologists had discovered unexplained layers of carbon lying next to dinosaur remains, and that they had also found unexplained deposits of radioactive materials lying next to dinosaur remains

I learned that entirely new kinds of plants and animals appeared in the geologic record almost immediately after extinctions—time after time after time. The new plants and animals had arrived as if from nowhere, with no known ancestors, with no intermediate life forms to explain their sudden presence.

Where did all of those new plants and animals come from? And why did their arrival so often coincide with geomagnetic reversals?

My curiosity was piqued.

How is it that evolution is supposed to work? One tiny step at a time? C'mon, I thought, let's get real. You don't cross huge chasms with timid little steps, you take one giant leap. Small steps are useless. What good is half an ear? What good is half an eye?

Then one day I ran across the answer.

"What good is half a wing?" asked geneticist Richard Goldschmidt in his 1940 book *The Material Basis of Evolution*. "Or half a jaw?" If evolution works in tiny imperceptible steps, asked Goldschmidt, why do we never find those intermediate stages in the fossil record?

Because they never existed.

New species do *not* evolve slowly, said Goldschmidt, they arise abruptly.

But why?

I knew what I had to do. I had to continue my research until I could answer that question – and as many of the other geologic questions as I could – point by point, in plain, everyday language. And thus was born *Magnetic Reversals and Evolutionary Leaps*.

At the time I made that decision, I didn't realize how acrimonious the debate over evolution can become.

It's the "Creeps" versus the "Jerks." The creeps (so dubbed by the jerks) insist that evolution creeps along in a smooth, gradual process occurring over vast periods of time. The jerks (so dubbed by the creeps), maintain that evolution works suddenly; that it makes sudden leaps.

Overall, I agree with the "jerks." But in my opinion, they have never presented a credible mechanism to explain their position—to explain what causes those leaps.

In these pages you will find such a mechanism. You will learn how geomagnetic reversals triggered those evolutionary jumps; you will learn that those jumps recurred according to a predictable, natural cycle, and you will learn that the next beat of that cycle is now due.

I hope you enjoy reading *Magnetic Reversals and Evolutionary Leaps* as much as I enjoyed researching and writing it.

* * *

Note to my previous readers:

I had originally included some of this writing in the first working version of *Not by Fire but by Ice*. In fact, the book had already gone through the editing process and was set to go to the printers. But at the last moment I decided that I was trying to cover too much territory, and pulled those words out.

Over the years I've added to my original research, but you still need some of the information found in *Not by Fire but by Ice* in order to make sense of what I'm saying. I therefore hope you'll understand when you see some of those earlier words repeated here.

See author's website at:
www.iceagenow.com

To my wife Pattie,
who brings sunshine, love and laughter
into my life

1

·····

SUDDENLY

·····

Remember Charles Darwin? He's the one who told us that evolution is a slow, stately and orderly process, plodding along so slowly that no one could hope to see it work in their lifetime. Darwin's theories of gradual evolution and natural selection are now regurgitated almost unthinkingly the world over.

    That's too bad . . . because Darwin was wrong.

"The idea of slow, gradual, imperceptible evolution is wrong," says paleontologist Robert Bakker, "the fossil facts do not read that way." Mass extinctions always come suddenly with no warning, says Bakker. The population is then immediately replaced with entirely different species with no forerunners.

"It's the best kept trade secret of paleontology," wrote Stephen Jay Gould of Harvard. "The evolutionary trees that adorn our textbooks have data only at the tips and nodes of their branches; the rest is inference, however reasonable, not the evidence of fossils."

Evolution works suddenly

Evolution works suddenly, said English paleontologist Hugh Falconer just one year after publication of Darwin's *On the Origin of Species* in 1859. "Most of the time," said Falconer, "species remain unchanged during the long intervals between sudden transmutation."

Darwin's friend, Thomas Huxley, didn't believe in gradual evolution either. Evolution moves so fast, said Huxley, that the slow process of erosion and sedimentation rarely catches it in the act. New species appear suddenly, with no intermediate fossils linking them to any known ancestor.

"You have loaded yourself with unnecessary difficulty in adopting *Natura non facit saltum* so unreservedly," said Huxley. (Nature does not make leaps.) After Huxley, anyone who dared entertain such a thought—that nature does make leaps—was labeled a saltationist.

But Darwin wouldn't budge. "The geologic record is exceedingly imperfect," said Darwin. "It's a book with few remaining pages, few lines on each page, and few words on each line. We do not see any intermediate change in the fossil record, not because change was abrupt, but because the intermediate steps are missing."

Frustrated paleontologists keep taking potshots at Darwin's theories, but it's like shooting into fog.

Take the 1972 paper *Punctuated Equilibria* by Niles Eldredge and Stephen Jay Gould. Most evolutionary change occurs suddenly, they said, like a punctuation mark in a sentence. Animals usually just sit there, not evolving at all.

Nature does take leaps

"Most new species appear with a bang, not a protracted crescendo," said Gould. "Gradualism is not a fact of nature. A species seems to remain unchanged in the fossil record for millions of years, before abruptly disappearing, only to be replaced just as rapidly with a species that is, though clearly related, substantially different. Nature does take leaps."

Why keep hammering on this? Because if evolution does take leaps, there must be a reason. That reason, you will soon see, can be linked to geomagnetic reversals and excursions.

Let's look at the record.

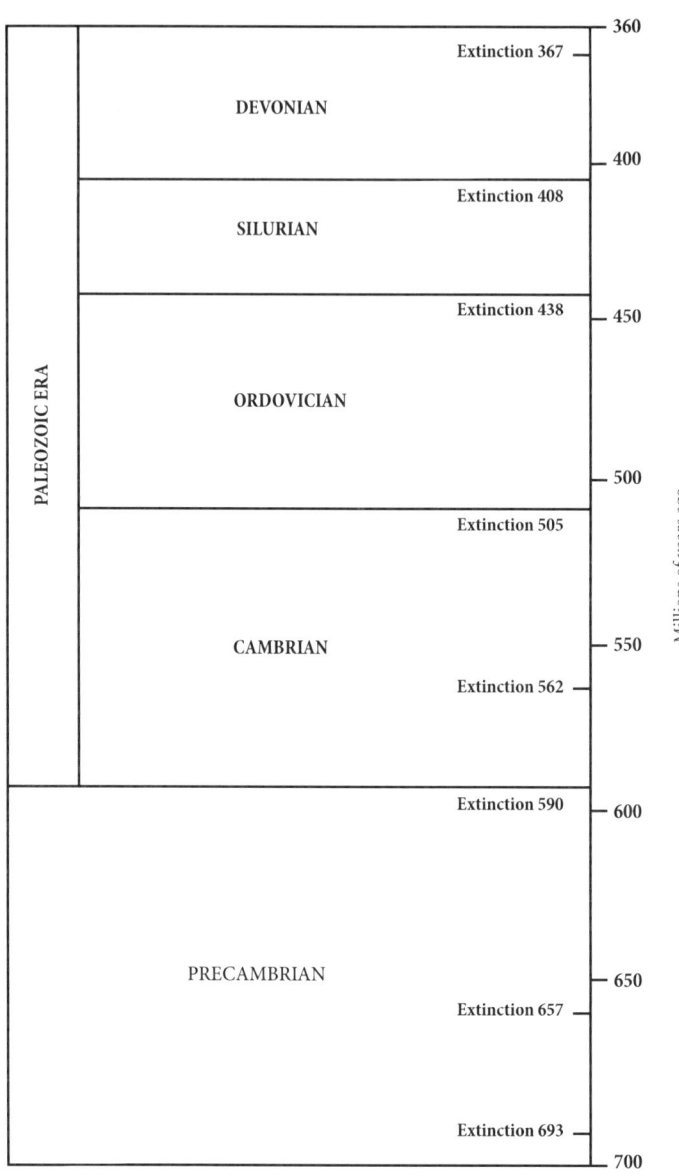

Geologic time scale: This page - 700 million years ago (mya) to 360 mya. Facing page - 360 mya to present.

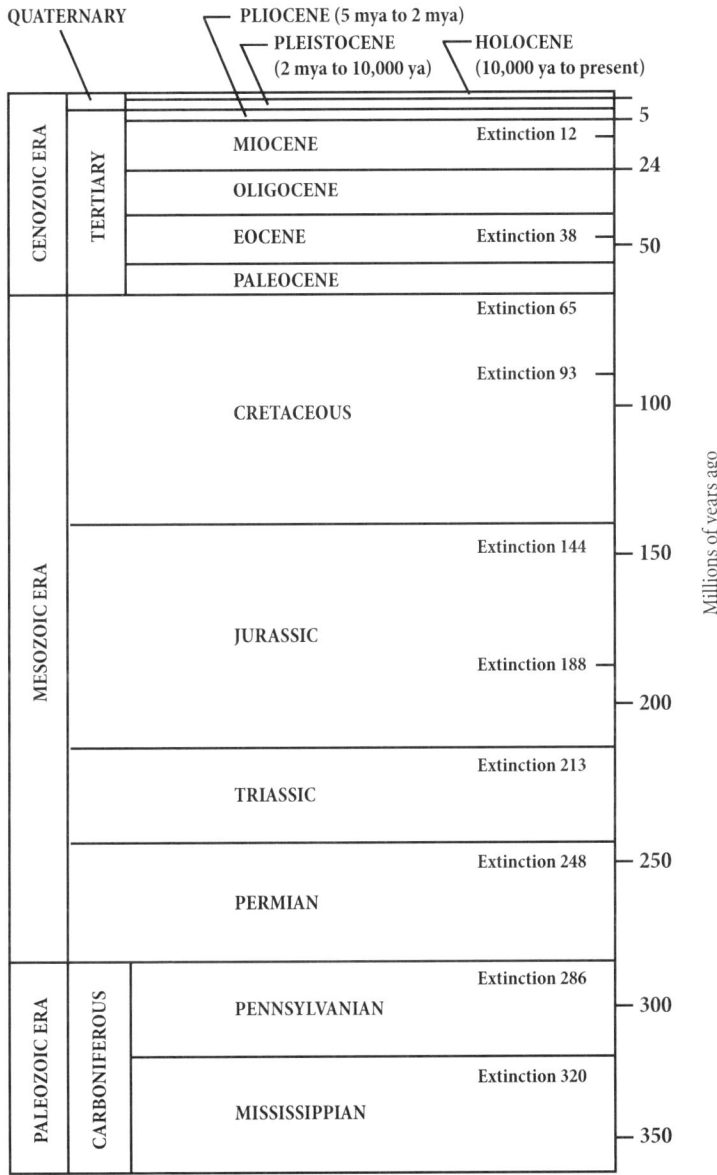

During the first three-and-a-half-billion years of life on our planet, earth's inhabitants were little more than connected chains of atoms floating around in a primordial soup of amino acids and other carbon-based chemicals. "Scum," one scientist called them. "Smears," said another.

Transforming themselves into jellyfish-like animals with no bones and no shells, the smears evolved into "squishy blobs" with no resemblance to any known organism dead or alive. Squishy blobs and algal mats (thin layers of prokaryotic algae which trap and bind sediment) were the highest forms of life on earth.

Did any extinctions occur during those billions of years? Almost certainly. But how do you find a squishy blob encased in a rock more than a billion years old? With no hard parts, such blobs almost never appear in the record. The first extinction that we're sure of, therefore, which apparently victimized only algae, took place about 650 million years ago.

The Cambrian Explosion

For all intents and purposes, though, life on earth began about 580 million years ago at the beginning of the Cambrian Period, when new life suddenly exploded across the earth. A dizzying array of new organisms appeared in the geologic record as if from nowhere. An astonishing burst of shelled forms took place, and virtually all major forms of animal life, including the ubiquitous trilobite, suddenly appeared. (Trilobites were distant cousins of today's horseshoe crab.)

Called the Cambrian explosion, the abrupt change from simple life-forms to more advanced was a critical turning point in the history of life. No one knows how or why it happened. What we do know is that the pattern has always been the same—sudden—always sudden.

Look at the Ordovician, when jawless fish with no known ancestors suddenly appeared; or the Silurian, when algae crawled out of the sea and onto the barren ground; or the Devonian, when coniferous trees suddenly appeared, as did ferns, seemingly out of thin air.

At the same time, ammonites spread through the seas, along with ever more bizarre types of giant armored fish sporting bony-plates on the outsides of their bodies.

Why haven't sharks evolved?

So many new kinds of fish appeared in the Devonian that it's called the "Age of the Fishes." Sharks appeared suddenly, and the first amphibian, *Ichthyostega,* crawled out of the water and onto the land. (Sharks pose a problem for Darwinian believers. Today's sharks are almost identical to those of 370 million years ago. If evolution is the only explanation, then why haven't sharks evolved into something "better"?)

Then, just as it looked as if the fishes had conquered the world, came yet another mass extinction leading to the Carboniferous. New kinds of ammonites made their debut, and insect life exploded across the planet. Fifteen new insect families (that's *families*, not *species*) suddenly arrived on the scene. And yet, there is no indication of their origin.

Insects similar to present-day cockroaches, grasshoppers, and mantids suddenly populated the globe, along with primitive forms of mayflies and dragonflies with wingspans approaching three feet.

And the cockroaches? Squash them with your shoe? Forget it. At 18 inches long, Carboniferous cockroaches were *longer* than your shoe. Cycads (similar to today's sago palms) popped out of the ground, and the first primitive reptiles suddenly appeared.

And don't forget the coal!

So much coal mysteriously appeared during the Carboniferous that we call it the "Coal Age."

Whatever happened to "slow, stately progression toward perfection"?

Happy is the man that findeth wisdom, and the man that getteth understanding.
— WORDS IN HYMN SUNG AT
CHARLES DARWIN'S FUNERAL

2

.....

LIFE OF A SOLDIER

.....

Whatever happened to "slow, stately progression toward perfection"?

There's no such thing. Geologic change *always* comes in spurts, said British geologist Derek V. Ager. "The history of any one part of the earth, like the life of a soldier, consists of long periods of boredom and short periods of terror."

Is there a gap in the geologic record? No, says paleontologist Kenneth Hsü, blasting Darwinian theories, there is no gap. "Many groups appeared suddenly, not because of the incompleteness of the geologic record, but because they *did* appear suddenly."

New kinds of ammonites

Following the Carboniferous extinction came the Great Permian, the granddaddy of all extinctions. Ammonites went reeling (only two families survived), and early reptiles took a beating. Trilobites completely disappeared, as did graptolites, vanishing with no known descendants. (Graptolites were sea-dwelling animals that formed colonies on floating stalks.)

Even insects. Eight of the twenty-seven orders of insects went extinct at the Great Permian extinction (Labandeira and Sepkoski, *Science*, 16 Jul 1993).

Drastically different kinds of life quickly invaded the seas. Like fresh troops sent into battle to relieve combat-weary veterans, entirely new kinds of ammonites (ceratids) suddenly appeared. They enjoyed an explosive radiation in the Triassic, only to be almost completely decimated at the end of the period.

Then came the diminutive thecodont, a smallish animal resembling a big pheasant, which strutted around on two bird-like legs. Who would have dreamed that these insignificant little creatures would be the forerunners of the great Jurassic dinosaur?

After yet one more extinction in the mid-Triassic, turtles and crocodiles suddenly appeared. Then—pow!—another mass murder, and the curtain fell on the Triassic. All thecodont families, and there had been many, went extinct.

When the curtain rose again, the world was an empty stage. But a new cast of characters quickly jumped into the limelight. Frogs, toads, and salamanders suddenly appeared with no identifiable ancestors, and entirely new kinds of ferns developed almost overnight.

As for dinosaurs, it was "the Golden Age of Giants," said paleontologist Robert Bakker, "as hordes of dinosaurs burst out of their Triassic bonds, and wave after wave of ever-larger species filled the land."

But did they evolve?

A resounding "no."

Twiddling their evolutionary thumbs

Sifting through a 250-foot-layer of the Morrison Formation at Como Bluff, Wyoming, Bakker dug through strata representing well over one million years. He followed *Brontosaurus* through hundreds of thousands of breeding cycles and through major climatic changes. "There was absolutely no evidence for continuous evolutionary change," says Bakker." They just sat there, "twiddling their evolutionary thumbs."

When they finally changed, they did it incredibly fast, growing 10% bigger all at once. *Allosaurus* impressed Bakker even more, ballooning in size by almost 50%. Then came another mass murder, and the first great "Age of Dinosaurs" came to an end.

But the battered dinosaurs would claw their way back to dominion one last time during the Cretaceous. Once again, entirely new kinds of ammonites fought their way to prominence. Fish-eating, sea-going lizards suddenly appeared, and sea turtles almost as big as a Volkswagen swam onto the scene. Where in the world did they come from? No one knows.

Abominable angiosperms

And where did the flowers come from? Angiosperms (flowering plants and hardwood trees) appeared immediately following the end-Jurassic. Enjoying an explosion of diversity, they quickly formed great forests of deciduous broad-leafed trees across the land, becoming the dominant vegetation and the main fodder for vegetarian dinosaurs. And yet, flowering plants had never before existed on the face of the earth. (Poor Darwin. This one had him buffaloed. He considered the sudden arrival of the angiosperms an "abominable mystery.")

Marine life expanded too, as pelagic carbonate organisms (floating ocean animals) burst out in an immense biological explosion all over the world. An invisible hand of creation, say paleontologists, *must* have been at work.

Rudistids were common. (The reef builders of their time, the rudist clam looked like an oyster pretending to be a coral, said paleontologist David Raup.) Other ocean dwellers, such as bryzoans, brachiopods, and foraminifera (plankton) became healthy, numerous, and vigorous. There is nothing—*nothing*—in the fossil record to indicate that their end was near.

Then disaster struck yet again, with more than seventy percent of all sea-dwelling animals going extinct at the end of the Cretaceous Period.

Whatever hopes the dinosaurs had of surviving were shattered as the relentless motor of extinction surged around the globe. Ammonites were simply massacred, and insects, again, almost annihilated.

The Fern Spike

North America's huge forests disappeared in less than one year, with nearly every kind of plant from Siberia to Alaska going extinct or becoming almost totally different.

Angiosperms disappeared almost overnight. Magnolia trees, or something similar, vanished in a single growing season; in four to six weeks, said paleobotanist Jack Wolfe of the U.S. Geological Survey (USGS). The only things that survived were the ferns. Fern spores suddenly made up almost all of the assemblage. It's called "the fern spike," and it proves that the extinction occurred suddenly. "It was an ecological St. Valentine's Day Massacre," said Stephen Jay Gould of Harvard, "the ecosystem totally collapsed."

Think of the disaster we would face today if angiosperms were to disappear. Almost everything that we and our mammalian relatives eat is an angiosperm, including corn, wheat, tomatoes, onions, cucumbers, squash, beans, and most of the berry-producing bushes and shrubs. Leeks, broccoli, and potatoes are angiosperms, as are maples, oranges, peaches, pecans and all of the other broad-leafed trees. More than 257,000 kinds of angiosperms exist on our planet.

How do you feed a fifty-ton dinosaur body when all of the food is gone?

For that matter, how do you feed a one-hundred-pound human body when all of the food is gone?

Not only at the end-Cretaceous, *every* geological era shows dramatic changes in plant life at the transition from one period to the next, with many plant forms completely disappearing.

Why didn't the seeds simply sprout back up in the same old ways as before? Some seeds *must* have been buried in the ground. What about the roots? Have you ever tried to eradicate blackberry bushes? Have you ever tried to eradicate dandelions? Bermuda grass? Don't blame forest fires, either. Most plants make fabulous comebacks after a fire. Did something happen to change their genetic make-up? The answer has to be yes.

No known ancestors

Immediately following the end-Cretaceous, primitive birds enjoyed a surge of development, and large flightless birds like today's ostrich suddenly appeared. Tree sloths, armadillos and anteaters saw dramatic development, as did egg-laying mammals such as the platypus and the echidnor. Other mammals, until then not much bigger than a house cat, took off in a rapid and spectacular diversification, all with no known ancestors.

Sea-dwellers enjoyed the same fate. Twenty-one of the twenty-seven species of lampshells (brachiopods) were completely obliterated at the K-T boundary, only to be suddenly replaced by twenty-four entirely new species.

Doesn't it seem odd that about the same number of new species would replace the old? "Though totally different," says David Raup, "the number of species has remained remarkably constant for almost 600 million years."

Gordian knot of geology

The new animals' shells were shaped drastically different from the old, and yet, there had been no time for evolution. It's one of the Gordian knots of geology, said William Berggren of the Woods Hole Oceanographic Institute in Woods Hole, Massachusetts. The new lampshells were so different that it "should" have taken considerable time for their evolutionary development.

But that's not how the record reads. The lampshells changed almost immediately.

Same with foraminiferas (forams).

Forams had evolved rapidly, and were at the zenith of their development, said Berggren. There was no sign of senility, no thinning of their ranks, until suddenly, "at a boundary so sharp you can mark it with a razor," all Cretaceous forms disappear. Sediments of the two periods are separated only by the fish clay, which contains no foram fossils.

A half-inch-thick layer of clay

Same in the Gubbio limestones.

Forams had been so prolific that their skeletons, combined with the calcareous (chalky) secretions of tiny plants known as nanoplankton, make up the bulk of end-Cretaceous sediments. Directly above the limestone is a layer of barren clay less than half-an-inch thick. Like Denmark's fish clay, the Gubbio clay looks like the sediment of a sterile ocean, with almost no kinds of fossil. On top of the clay is limestone again. But this time the limestone is made of extremely small forams. Microscopic, the new forams are only one-tenth the size of the ones they replace.

Whoa! One-tenth the size? If that should happen to us, the tallest basketball player in the world would be nine inches tall. Shaquille O'Neil—the size of a Barbie Doll? That's a *drastic* change!

According to paleontologist Gerta Keller at Princeton University, some of those new forms were only one-twentieth the size of the old ones.

Mike Tyson—the size of an angry mouse standing on its hind legs?

What time is there for evolution when the only thing that separates the new fossils from the old is a thin layer of clay? A half-inch-thick layer of clay could be deposited in a matter of months.

Or days.

Or in one stormy afternoon.

Other new animals also appeared. Hoofed mammals, such as deer, horses, cattle, and sheep galloped onto the scene, while spanking-new marine mammals, including sea lions, baleen whales, toothed whales, seals, small porpoises, and dolphins joined the escalating "evolutionary" battle. Nose-twitching, fur-bearing, mouse-like mammals scampered across the land, and modern bird orders along with mammalian bats suddenly flew through the skies.

After yet another mass extinction about 37 million years ago, gorillas, chimpanzees, and monkeys swung into view; hens and rabbits suddenly appeared; and giant ground sloths bigger than an elephant lumbered into the picture. Pigs, bears, and the first true dogs and cats elbowed their way into the ranks, along with a giant no-horned rhinoceros standing two stories tall at the shoulder. They all arrived suddenly, with no similarity to the animals they had replaced.

The theory of survival-of-the-fittest calls for competition, with the best "man" winning. But there was no competition, no battle, no culling out of the sick or old, no superior forms coming along to shove aside inferior or incapable forebears. There was nothing about the new arrivals, scientists insist, that made them any better than the old ones.

Newcomers arrived fully developed

"We have no evidence whatsoever—not a shred—that losers in the great decimation were systematically inferior in adaptive design to those that survived," said Stephen Jay Gould. "None of these ancient groups shows any sign of anatomical insufficiency." The new species simply arrived, fully developed, tossed into the ring with no fanfare.
Fully developed!
How is a poor, bewildered, "just-the-facts-sir-just-the-facts" scientist supposed to explain it? It's as if the very act of extinction somehow leads to entirely new forms of life. It's more akin to the Biblical sense of creation than to Darwinian evolution.

The grim reaper kept swinging his deadly scythe, making yet another seventy-percent sweep just two million years ago at the beginning of the Pleistocene, and an entirely new batch of animals suddenly appeared.

True elephants and hippopotamuses tromped across the land. Zebrine horses and antelope bounded into the fray, along with musk-oxen, beaver, reindeer, lemmings, and foxes, followed by woolly-rhinoceroses, woolly-mammoths, mastodons, moose, sabre-tooth cats, and the great dire wolf.

Did you know that all of those animals arrived suddenly within the last two million years, with no time to evolve? Among the many species of mammals now existing in Europe and Asia, all but six appeared during the past two million years, with no time to evolve.

Did you know that we (anatomically modern humans) are blindingly new?—that we've existed for only 200,000 years or so? (Twenty years ago, scientists thought we had existed for only 50,000 years.)

Two hundred thousand years. "And we have not a shred of evidence for any genetic improvement since then," said Stephen Jay Gould. "I suspect that the average Cro-Magnon, properly trained, could have handled computers with the best of us. And for what it's worth," Gould added, "they had slightly larger brains than we do. We've grown from perhaps one hundred thousand people with axes to more than four billion with bombs, rocket ships, cities, televisions, and computers—and all without substantial genetic change."

Where is evolution?

Bigger animals savaged

Finally, 11,500 years ago at the end of the last ice age, history repeated itself yet again. And again, the larger animals were savaged, with some seventy percent of the megafauna taking it on the chin. Mammoths, mastodons, sabre-toothed cats, along with 40 million other animals, and possibly much of the human race, disappeared in a geologic instant

But for some reason, the same animals and plants that died in the north continued to thrive in the south. For some reason, the killer showed more compassion in Africa and South America.

So what does this tell us?

It's high time that we admit that the Darwinian theory of slow, stately evolution is a myth. The changes come fast and furious, and almost always arrive on the heels of a mass extinction.

Savants and priests of the earliest cultures knew that the earth was flat. To think otherwise was absurd. If the earth had another side beneath our feet, rain would have to fall upwards, water would not stay in lakes, and people would have to walk upside down.
— S. WARREN CAREY

3

.....

A DEADLY MISTAKE

.....

Mass extinctions have been the rule, rather than the exception, for the 3.5 billion years that life has existed on this planet. Almost identical, each extinction was abrupt, each was extensive, and each was caused by some temporary, unexplainable event. What could that temporary, unexplainable event have been?

For clues, let's look at one of the most famous extinctions of all time, the dinosaur extinction of sixty-five million years ago.

Seventy-five percent of all species go extinct

Calling it the "dinosaur extinction" is misleading, because dinosaurs weren't the only animals that died.

It was mass extinction, global and sudden. Seventy-five percent of all species on the planet went extinct, never again to appear in the geologic record. The sheer number of other deaths, say scientists, make the dinosaurs' disappearance look almost like an afterthought.

The dinosaurs' graves mark the end of one geologic period and the start of another; the end of the Cretaceous Period and the beginning of the Tertiary. It's known as the K-T boundary.[1]

And what happened to the plants? Though most vegetation at the end-Cretaceous consisted of flowering plants and trees, tree pollen almost totally disappeared.

---

[1] The 65-million-year-old dividing line between the Cretaceous and Tertiary periods is called the K-T boundary because the first Cretaceous fossils were found buried in chalk. "Chalk" in Latin is *creta*, hence the name Cretaceous. The "K" comes from *kriede*, the German word for *creta*, and the "T" stands for Tertiary, which means third. The K-T boundary therefore marks the beginning of the third major era in geologic history.

Tree pollen almost disappeared

Plant pollen and spores disappeared so rapidly at the K-T boundary, said Robert Tschudy of the U.S. Geological Survey, that "we can envision the forest dead in less than a year's time. The catastrophic change, Tschudy said, compares to the scenario after a volcanic eruption as ferns establish tenuous new footholds in barren fields of lava.

Meanwhile, many world-class geologists contend that volcanoes killed the dinosaurs. It's a legitimate argument. Massive volcanic eruptions occurred all over the world right at the K-T boundary.

Kevin Padian, curator of the Museum of Paleontology at the University of California, Berkeley, adds fire to the volcanic theory.

Padian, along with Cynthia Faux, a paleontologist at the Museum of the Rockies in Bozeman, Montana, points out that many dinosaur bodies have been found twisted and contorted, with their mouths wide open, heads thrown back and tails recurved, as if the beasts had suffered violently at death – perhaps asphyxiated by volcanic gases or ashfalls. Many dinosaurs, ranging from the flying pterosaurs to *Tyrannosaurus rex*, as well as many early mammals, exhibit similar agonized postures.

Sea dwellers almost decimated

Sea dwellers big and small, from giant sea-serpents to rudistid clams to microscopic plankton, were almost decimated.

But why?

Sea dwellers had not declined or lost vigor, said Surlyk and Johansen in *Science*. Healthy and numerous, there was no warning of their impending doom, neither in the form of decreasing population density nor in diversity.

And why were certain animals spared?

Why were ocean dwellers almost annihilated, while freshwater animals survived? Crocodiles and alligators came through with flying colors, as did freshwater turtles.

Fourteen percent or less of freshwater dwellers died, while freshwater mollusks underwent no significant deaths at all, says Charles Drake of Dartmouth College. What kind of mad killer would destroy seventy percent of sea dwellers while killing only fourteen percent of freshwater inhabitants?

Why did ocean dwellers who secreted calcium carbonate almost disappear, while those that secreted silica remain almost untouched?

Ocean temperatures shot upward

And why did ocean temperatures shoot upward? Ocean temperatures went *up*, mind you, not down. Paleontologist Kenneth J. Hsü (1982) thinks ocean temperatures shot upward an incredible $10°$-$12°C$ ($18°$-$22°F$). The increase was abrupt and short-lived.

What could have heated all of the world's oceans by eighteen to twenty-two degrees?

For that matter, what could have pulverized Denmark's fish? That's right, pulverized their fish! Denmark's boundary clays contain so many shattered fish fragments that they're called the fish clays (the *fiskeler*).

The fish clays, on the coast south of Copenhagen near the town of Stevn's Klint, consist of a four-inch-thick layer of dark-gray clay sandwiched between a layer of white chalk below, and two layers of white limestone above. Except for the broken bones, the fish clays contain almost no hint of life.

Black muds

Not only acidic, end-Cretaceous seas were also anoxic (deficient in oxygen). Scientists assume that the seas were anoxic because of the black muds that settled to the bottom of the seas; black muds that later became black shales, black shales on top of apparently healthy reefs.

"The sea suffered from indigestion," said Alfred Fischer of Princeton University. So much carbon and soot poured into the seas that they couldn't metabolize it all.

Again, it's a recurring thing.

Carbon-laced black shales are found at the ends of several geologic periods including the Cambrian, the Permian, the Silurian, the Ordovician, the Devonian, and the Late Triassic. Digby McLaren at the University of Ottawa speaks of black shales at the Frasnian-Famennian extinction while Anthony Hallam tells of black shales at the end-Aptian, end-Pleinsbachian, and end-Cenomanian. Where did that carbon come from?

Carbon anomalies also occurred on land. There's so much carbon in the boundary clays, said University of Chicago graduate student Wendy Wolbach, that more than ninety percent of all vegetation on earth, and many millions of animals, must have burned as raging forest fires engulfed the planet.

Black and dirty, soot-filled skies would have turned day into night. The darkness, combined with frigid temperatures could have lasted for months, if not years.

And what about the coal? Many dinosaur bones are found lying beneath a layer of coal. Sometimes entire dinosaur skeletons are found lying within the coal itself. (More about dinosaur bones and coal later.)

Larger animals at risk

Those are just some of the mysteries. On land, all animals weighing more than 55 pounds were killed, while many small animals survived. Many families of lizards and mammals passed through the disaster almost unfazed.

So what's the answer?

When paleontologists try to understand the past, they look at the layers of soil, clay, and rock - and the fossils embedded in those layers - that were deposited during the period they're studying.

In 1978, geologist Walter Alvarez made a serendipitous discovery in the Apennine Mountains near Gubbio, a small town in central Italy.

Alvarez had gone to Gubbio because he was interested in periods of reversed polarity; times when compasses would have pointed south instead of north.[1]

Iridium

As Alvarez conducted experiments on the Gubbio clays, he found abnormally high concentrations—up to 30 times greater than normal—of iridium. Iridium is a metallic element resembling platinum. He also found unusually high concentrations of osmium, antimony, and arsenic. (For what it's worth, Denmark's fish clays contain 200 times the normal amount of iridium.)

Would he find iridium in other locations?

A resounding yes!

Alvarez found a half-inch-thick layer of iridium-laced clay all across the globe right at the K-T boundary. Boundary clays from around the world, it made no difference where they came from, all contained excess iridium. Boundary clays from New Zealand held 20 times normal concentrations. Boundary clays from Denmark held even more.

---

[1] Compasses would have pointed south during roughly half of history, said Allan Cox and Robert Hart in their 1986 book *Plate Tectonics, How it Works*. The earth's polarity has reversed hundreds, perhaps thousands, of times. The K-T extinction occurred near one of those reversals.

But how did such huge amounts of iridium get scattered across the planet? Though there seems to be plenty of iridium in the earth's core, it's rare on the crust. How did it get to the surface? And in a layer? It looked as if it had come from the sky.

It did come from the sky!

Iridium is relatively common in celestial bodies such as asteroids and meteorites, Alvarez recalled. The iridium points to a cataclysmic collision between the earth and a giant galactic intruder, he proclaimed. A mountain-sized asteroid at least six miles in diameter had crashed into the earth head-on.

Reversed magnetic polarity

But wait. What about that reversed polarity? After all, wasn't that why Alvarez went to Gubbio to begin with? Could a magnetic reversal have caused the extinction? No, most scientists insist, a polarity reversal would have caused "no meaningful consequences." The earth's magnetic field has reversed itself many times in the past, they point out, with no ill effects (that they know of).

But they have to say something, so they usually toss in an aside about some guy named Uffen, who came up with a magnetic reversal theory back in the 1960s. Our magnetic field shields us from cosmic rays, said Uffen. When the field reversed we temporarily lost our shielding, and mutation-causing cosmic rays bombarded the earth.

Cosmic rays? Far out. No one paid much attention to Uffen.

What a deadly mistake.

It would be an irony if one of the elements that led Alvarez to the planetesimal cause of the extinctions was itself the agent of those extinctions.
—MICHAEL ALLABY & JAMES LOVELOCK

4

......

THE IRONY

......

Forty precious years ago, Robert Uffen warned us about magnetic reversals; forty precious years, during which we could have been preparing for the coming disaster. Now we need to play catch-up, and we need to do it fast.

Our magnetic field (the magnetosphere) shields us from the charged particles in the solar winds, said Uffen, of

Canada's University of Western Ontario. During a geomagnetic reversal, cosmic radiation would bombard our planet, leading to mutation or death.

I think Uffen was right.

So do many others.

New kinds of animals appear in the geologic record "virtually simultaneously" with magnetic reversals, said Kennett and Watkins of the University of Rhode Island.

"Reversals strongly influence population trends," said C. J. Waddington at the University of Minnesota.

"Faunal (animal) changes occurred near several reversals," said Allan Cox of Stanford University.

Radiation and mutation

But why? It all goes back to radioactivity.

"We have long known that exposure to ionizing radiation increases mutation rates," said paleontologist John F. Simpson. During a polarity reversal, the earth's magnetic field strength would drop to zero, allowing excess radiation into our skies. Geomagnetic field strength is decreasing right now, Simpson noted.

But other scientists disagreed.

It would make no difference if magnetic field strength dropped all the way to zero, they said. Even if a charged particle weren't deflected by the magnetic field, it would need to be traveling at an almost impossibly high rate of speed to make it through our atmosphere and to the ground. (Some cosmic particles do travel at impossibly high rates of speed. More later.)

Today, the idea that radioactive materials could rain to the earth is considered a fact. Radioactive elements, such as carbon-14, are constantly created high in our skies when the speeding particles in cosmic rays collide with nitrogen atoms in the atmosphere. This process, called nuclear electron capture, adds or removes electrons from atoms or molecules that were previously neutral. Add a neutron to almost any atom, said Willard F. Libby, onetime commissioner of the Atomic Energy Commission, and it will become heavier and frequently radioactive.

Beryllium-10 is another radioactive element created high in the sky. So are helium-3 and tritium. Small amounts of radioactivity are falling to the earth this very second. That radioactivity can be linked to our magnetic field, and to geomagnetic reversals.

Geomagnetic reversals

Peel back the layers of mystery surrounding polarity reversals, peel back the layers of clay at Gubbio, and you'll solve the entire extinction enigma.
You'll also destroy the idea of slow, gradual evolution.

Let's begin peeling.
How can we tell that our magnetic field was reversed at a particular time in history? Through magnetostratigraphy, the study of the magnetic properties of ancient layers of sediment (strata) now hardened into rock.
Magnetic materials, such as magnetite, occur in all rocks. Like miniature magnets, they're tiny pieces of

ferrous metal that aligned with the earth's magnetic field as the rocks were being formed. As sedimentary grains drift to the bottom, the earth's magnetic field twists them in the water (like compass needles) until they align with the field. By determining their lie—which way they point—scientists can tell which way was north at the time the strata formed.

Magnetism and the Curie temperature

So too with igneous rocks and basalt. Magnetites in magma and lava also align with the earth's magnetic field. Non-magnetic while hot, their iron and titanium oxides become magnetic as they harden and cool through the Curie temperature, thus writing a record in stone as to when the rocks were created.

Those rocks show that our magnetic field has reversed itself many – probably many thousands – of times throughout history.

Periods of reversed or normal polarity are divided into magnetostratigraphic epochs. To honor the work of the early pioneers, today's epoch, which began about 780,000 years ago at the Brunhes/Matuyama boundary, is called the Brunhes Epoch. The epoch before today's, an epoch of reversed polarity, is named for Matuyama.

Magnetic excursions aborted reversals?

In addition to full-scale reversals, our magnetic field sometimes moves away from north for short periods of time, and then moves back. These movements, called magnetic excursions, are found in lava flows in many parts of the world, and from many different periods.

Magnetic excursions usually began suddenly as the magnetic North Pole moved rapidly and smoothly toward the equator. Sometimes it popped back to its original position almost immediately. Other times it crossed the equator and moved part way through the opposite hemisphere before swinging back to its near axial north-south position. Magnetic excursions, many paleomagnetists believe, were actually aborted reversals.

Geomagnetic excursions are generally brief, ranging from about five hundred years to perhaps five thousand years. (Mankinen and Wentworth, 2003)

Because excursions can be so short-lived, they can also be easy to miss in the geologic record, and can be easily confused with other excursions or reversals.

There's still a lot to learn about magnetic reversals, but one of the most exciting discoveries is that they may recur in a pattern. Not only in a pattern, magnetic reversals appear to occur in sync with ice ages and with precession of the equinoxes.

Why is this important? Because mutation-causing radiogenic materials bombard our planet in sync with the same cycle.

And it happens *fast!*

Strontium

Evidence comes from the rapidly changing strontium ratios in the seas. The changes show no lag time, said Philip Froelich in *Nature,* the ratios change in phase with the growth and decay of the ice. But that's impossible, said Froelich, "strontium has a residence time in the sea of several million years, too long to explain such rapid rates of change."

Strontium ratios change so rapidly, said Clemens, Farrell, and Gromet of Brown University, that "there may be glacial-age strontium sources not yet accounted for." (You bet there are strontium sources not yet accounted for. More later.)

Dramatic spikes in radioactivity

Dramatic spikes in other elements also appear in the record. Radioactive carbon-14 levels increased by 300% to 400% at the end of the last ice age, said the French scientist Alain Mazaud. This strongly reinforces the hypothesis, said Mazaud, that geomagnetic variations are the major source of long-term variations in the abundance of carbon-14.

Another drastic increase in carbon-14 (four to five times normal) occurred around 22,000 years ago, said Edouard Bard of Lamont-Doherty Earth Observatory. The most reasonable explanation, said Bard, "involves magnetic modulation."

Carbon dioxide levels also increased (by about thirty percent) at the end of the last ice age (Neftel *et al.*). And again, it wasn't the only time. The carbon dioxide record exhibits a cyclic change corresponding to precession, said J. M. Barnola. It's called the carbon cycle.

Beryllium-10

Same with beryllium-10. Twice the normal amount of beryllium-10 is found in 60,000-year-old ice, said G. M. Raisbeck of the Laboratoire René Bernas in Orsay, France. Another spike in beryllim-10 occurred about 35,000 years ago. "The dramatic difference may have been caused by changes in magnetic intensity," said Raisbeck, "and may be related to the Lake Mungo magnetic excursion." (More about Lake Mungo later.)
Another spike in beryllium-10 occurred 780,000 years ago at the Brunhes magnetic reversal. Other spikes occurred at 105,000, 90,000, 68,000, and 23,000 years ago (at the Mono Lake magnetic reversal). The most recent spike—two to three times normal—occurred about 11,000 years ago during the Gothenburg magnetic reversal (Beer, 1984).

Even methane. Methane levels doubled at the end of the last ice age. And again, it wasn't the only time. Changes in methane levels are cyclic, and can be linked to orbital variations (Chappellaz).
Peaks in sulfate, nitrate, and chloride also occurred at the end of the last ice age (Herron and Langway).

A layer of black carbon

Same with soot. While exploring ancient caves for his doctoral thesis, John S. Kopper of Columbia University discovered a layer of black carbon deposited near the end of the last ice age. It must be soot, Kopper theorized. Maybe early humans managed their crops by burning.

Sure, blame it on humans. That's who we tried to pin the mammoth extinction on, too.

But should we take the rap for all of the other debris swirling through those ice age skies?

No way.

Nor should we take the rap for today's increases in the same elements. Carbon dioxide levels, along with methane, hydrocarbons, sulfur and nitrogen oxides, and others, are increasing daily, said Dixy Lee Ray and Lou Guzzo in their 1990 book *Trashing the Planet*. The rate of increase is "substantial," about one percent per year.

Again, many scientists blame humans. "But it's not as simple as that," said Ray, former chairman of the Atomic Energy Commission. "Such increases have occurred in the past without any help from us at all, and this time is probably no different. Most likely, the causes were and still are colossal cosmic forces quite outside human ability to control."

Which brings us to geomagnetic reversals and equinoctial precession. Put those two forces together, and you've found that "colossal cosmic force" that Ray and Guzzo were talking about.

Think about it.

Layers of black soot; increases in carbon-14, strontium and beryllium-10; increases in methane, sulfate, nitrate, chloride, carbon dioxide, and hydrocarbon levels, all in sync with geomagnetic reversals—and we don't think magnetic reversals are important?

Even if fishes hone their adaptations to peaks of aquatic perfection, they will all die if the ponds dry up. But grubby old Buster the Lungfish, former laughing-stock of the piscene priesthood, may pull through—and not because a bunion on his great-grandfather's fin warned his ancestors about an impending comet.

—STEPHEN JAY GOULD

5

HALF A WING

Remember that old story of the tortoise and the hare, where the slow-moving tortoise wins the race? Mass extinctions work the same way. The race goes to the slowest at extinctions, not the fastest. It is not a crapshoot. It is not Lady Luck in reverse. At extinctions, the energetic animals, the active animals, the high-metabolism animals, are always the first to die.

They're also the first to "evolve," continually branching into ever more divergent species, while low-metabolism animals tend to remain unchanged. The very attributes that serve the active animals so well in everyday life defeat them in the clinch.

Top predators at risk

If "survival of the fittest" is the determining factor, then a high metabolic rate should ensure survival. The winners should be the fastest, meanest, smartest, most efficient, most ferocious animals around. But it doesn't work that way. Mass extinctions are especially unforgiving to top predators, said paleontologist Robert Bakker in his 1986 book *Dinosaur Heresies*.

Compare the speedy ammonite to the more sluggish nautilus. The ammonite deposited thousands, perhaps tens of thousands, of minuscule, quick-hatching eggs in shallow water, while the nautilus laid only a few eggs in water more than 300 feet deep, says Peter D. Ward, professor of geological sciences at the University of Washington. The eggs then took a year to develop.

During normal times the ammonite expanded like crazy, but at the K-T extinction its speed became its own death warrant. The nautilus's slow reproductive methods made the difference, Ward surmises.

Sluggards slipped across the boundary

Sluggards of all kinds slipped across the K-T boundary with almost no change. Many reptiles and amphibians, including alligators and crocodiles, survived handily, as did the snapping and soft-shelled turtles. Gill-breathing salamanders coasted through along with the monitor lizard.

Same with the clam. Most kinds of clams move around very little, and yet, they've muddled through life for millions of years with practically no change.

Or consider the giant tortoise. The size of a Volkswagen Beetle, the giant tortoise's snail's-pace evolution gives eloquent new meaning to the words "as slow as a turtle." Suddenly appearing during the Eocene about fifty million years ago (the only sudden thing it's ever done) the giant plant-eater evolved so slowly that all new tortoise species of the past five million years can be placed into a single genus.

Living in the fast lane

Now compare the tortoise's lackadaisical evolution to that of the duckbill dinosaur. First appearing some 80 million years ago, duckbills expanded at a breakneck speed for the next ten million years. They evolved so rapidly that seven different genera are found in one small outcrop of the Judith River Formation alone, said Robert Bakker. Their success was staggering. Why so vulnerable at the end?

How to avoid the duckbill's fate

What's the best way to avoid the duckbill's fate? - the best way to avoid extinction? Play dead, be lazy, and don't get too hungry or thirsty.
And what's the best way to ensure your death? Be endowed, says Bakker, with the highest, most compulsive need for calories and protein.
That's why animals who live in the fast lane get hammered so hard at extinctions. They're hungry. Forced to come out of hiding too soon, they're bombarded by radiation or eat radioactive food. Low-metabolism animals simply wait it out. The race is not to the swift.

But how does this apply to dinosaurs? Weren't dinosaurs merely a bunch of dim-witted, cold-blooded, slow-moving reptiles, stumbling through life at barely moronic levels? No, no, and no again. Dinosaurs weren't stupid at all. With a guesstimated IQ of 50, many were as bright as the average golden retriever or chimpanzee. They weren't laggards either. Some dinosaurs could run up to 50 miles an hour.
Nor were they reptiles.
Many kinds of dinosaurs, says Bakker, were birds—earth-bound flightless birds. Not everyone agrees, but Bakker dismisses them. "There are still a few of my colleagues," says Bakker, "who think if it walks like a duck, breathes like a duck and grows like a duck, it must be a turtle." (*Time*, 26 Apr 1993).

But birds eat prodigious amounts of food. If dinosaurs were indeed related to birds, the bigger ones must have required hundreds of pounds of food per day. They weren't prepared for a nuclear attack. They had no granaries or well-stocked food bunkers. Bombing raids or not, they had to eat. Dashing into the holocaust to seek food, they were double-whammied; first from the outside by radiation, then from the inside by contaminated food. (This is not just idle speculation. Unexplained deposits of uranium are often found near dinosaur digs.)

How is it that evolution is supposed to work? One tiny step at a time? C'mon, let's get real. You don't cross huge chasms with timid little steps, you take one giant leap. Small steps are useless. What good is half an ear? What good is half an eye?

What good is half a wing?

"What good is half a wing?" asked geneticist Richard Goldschmidt in his 1940 book *The Material Basis of Evolution*. "Or half a jaw?" If evolution works in tiny imperceptible steps, asked Goldschmidt, why do we never find those intermediate stages in the fossil record?
Because they never existed.
New species do *not* evolve slowly, said Goldschmidt, they arise abruptly, in a process he called macromutation. Most newcomers could only be viewed as disastrous, he believed, as "monsters." But every once in a while, by sheer good fortune, a "hopeful monster" somehow adapted itself and lived. The rare success of these hopeful

monsters, said Goldschmidt, not the accumulation of small changes, is the real motor of evolution.

In the 1940s, Goldschmidt's ideas were scorned. After all, what could cause a macromutation? But in today's age of genetic engineering, I doubt that many people on our planet don't understand what radiation can do. It alters the cells that contain DNA, causing genetic changes.

In hindsight, it seems so obvious: Put Uffen and Goldschmidt together, and you've found what drives "evolution." (Uffen was the one who thought radiation would flood our skies during a magnetic reversal.)

Suppose you get zapped by radiation. What next? Depends on how much you get. Four thousand rads and your nervous system will go haywire in minutes. You'll bleed internally, endure bouts of bloody diarrhea and vomiting, and your blood pressure will drop through the floor.

Wracked by convulsion-like epileptic seizures, your mental ability will disintegrate. Then you'll lapse into merciful unconsciousness.

Then you'll die—all in two to three days.

Mutant monsters

But what if you live?

If you aren't sterile, you may give birth to a mutant. No arms? No legs? Five eyes? No eyes? Who knows what kind of mutant monster may pop out. Will it have a horn sticking out of its nose? A leg sprouting from its head? How about a tail?

A tail? Don't laugh. Every once in a while, a modern hospital will report the birth of a human baby with an unmistakable tail, says paleontologist Robert Bakker. It will be a normal child with all of the expected organs. But when you pick it up, trailing out behind will be a caudal appendage protruding beyond the buttocks for two to three inches.

Not just a wispy hint of a tail, either. Some of the tails are bigger than the average caudal remnant on our closest kin, the chimps, gorillas, and orangutans, says Bakker. If such a thing can occur in today's world, imagine what a sky filled with radioactivity could do.

Actually, you don't need a sky-full.

Babies in the womb much more sensitive

Studies of pregnant women from the Hiroshima and Nagasaki areas show that babies in the womb are extraordinarily sensitive to radioactivity; much more sensitive than their mothers. One-hundred rads and the newborn may emerge with microcephaly (reduced brain size), or exencephaly (part of the brain outside the head), or hydrocephaly (enlargement of the head due to excessive fluid), or anophthalmia (improperly developed eyes). A dose as low as 25 rads can cause major problems with eye development. Fifty rads can cause mental retardation.

Other than nuclear war, how would an embryo become subjected to radiation? From X-rays, gamma rays, and ultraviolet rays. Ultraviolet rays don't penetrate deeply enough to affect humans, but gamma rays and X-rays do.

And where do we find those kinds of rays? In the same cosmic rays that bombard our planet 24 hours a day. They're electromagnetic waves, moving energy from one place to another.

Let that sink in for a moment. Cosmic rays, X-rays, gamma rays, and ultraviolet rays are electromagnetic waves.

How can we possibly believe that our magnetic field could jiggle around, or actually reverse, without affecting every electromagnetic wave on this planet? Twisting, turning, realigning with the reversing magnetic field, the potential for nuclear collisions—and thence radiation—must be enormous.

A single radiation event, no matter how small, can cause mutations. "There is no lower limit to the amount of radiation which will increase the number of mutations," said John F. Simpson in a 1966 *Geological Society of America Bulletin*. Any increase, said Simpson, no matter how slight, will increase mutation rates by some amount. Mutations can be caused by doses as low as eight rads.

That, I propose, is why the faster-hatching ammonite died. It emerged into a world still bathed in radioactivity. The lethargic nautilus, crawling into the world as much as a year later, escaped the brunt of the attack and survives to this day.

Man has been here 32,000 years. That it took a hundred million years to prepare the world for him is proof that that is what it was done for. I suppose it is. I dunno. If the Eiffel tower were now representing the world's age, the skin of paint on the pinnacle-knob at its summit would represent man's share of that age; & anybody would perceive that that skin was what the tower was built for. I reckon they would. I dunno.

—MARK TWAIN

6

....

OOPS

....

Radioactivity might solve a few other riddles, too. Bones and teeth contain elements of fairly high atomic weight, so they absorb much more radiation than those of low atomic weight. Since bone consists mainly of calcium phosphate, that enables radioactive strontium, which is chemically similar to calcium, to be easily deposited in the bone.

Maybe that's why bones, horns, teeth and tusks get elongated, compacted, fattened, flattened, multiplied, and altered the most. Maybe that's why the sea-dwellers who secreted calcium carbonate almost disappeared, while those that secreted silica (silicone dioxide) were barely touched.

It's not survival of the fittest. Mankind didn't evolve and neither did anything else.

Call it the hand of God, call it natural selection, call it mutation, call it what you like; we were placed on this earth just the way we are, fully formed and fully developed. And that's essentially the way we will stay, until we too get wiped from the slate.

We may grow bigger or we may grow smaller. But we will not change in any fundamental way until the hand of creation reaches down once again. It's the closest thing you'll ever see to the Biblical sense of creation.

It's not evolution, it's creation

It's not evolution, it's creation. It's not evolution, it's trial and error. Oops, sorry 'bout that, half a jaw doesn't work. Oops, legs on top of the head look pretty silly. Oops, it's hard to hear without ears. Oops, oops, and oops.

How can we possibly look at the weird ways that our bodies have been put together, stretched, squeezed, and sometimes taken apart again, and not help but wonder: Was there a purpose?

Look at the horse.

Today's horse has one toe (we call it a hoof). But 50 million years ago, it had four. Every now and then, a healthy mare with the "right" number of toes (one) gives birth to a foal that has additional little toes sticking out beside the hoof. What possible purpose can we ascribe to this phenomenon?

Or look at the whale.

Modern whales have no hind legs at all. But in earlier times, back when they were land-living predators, they did. Every once in a while, says paleontologist Robert Bakker, "a modern whale is hauled in with a hind leg, complete with thigh and knee muscles sticking out of its side. These atavistic hind limbs are nothing less than throwbacks to a totally pre-whale stage of their existence, some 50 million years ago."

Now, if we call it a great evolutionary advance that man evolved from a fish by growing legs, what do we call it when a whale discards those legs? An evolutionary retreat? Have we really gotten better? Or just different?

Dinosaurs just looked different

Just different, says Jack Horner of the Museum of the Rockies at Montana State University in Bozeman. "Dinosaurs basically aren't any different from animals alive today. They just looked different," says Horner.

Who could argue that dinosaurs didn't look bizarre? But who could persuade one of those five-story giants that a half-naked, no-horned, glasses-wearing, bald-headed, beer-guzzling, T-shirted, tiny human being isn't really the one who is weird?

Some dinosaurs tromped around on four pillar-like legs; others flitted about on two bird-like feet and laid shell-covered eggs on the ground. Yet other dinosaurs were flashers. Sauntering around the Cretaceous swamps with an eight-foot-tall spiny sail on its back, the Spinosaur must have looked like a walking sexual billboard.

Some dinosaurs ate plants, while *Tyrannosaurus rex* ate meat with a mouthful of teeth the size of pickaxes. Other dinosaurs had beaks. Eggs, beaks, and webbed feet. Good reasons to suspect that dinosaurs might be related to birds, don't you think?

Barosaurus could stretch its neck fifty feet into the air; a modern giraffe at full height can reach eighteen. And *Apatosaurus*, at 55 tons, weighed as much as an entire herd of elephants. The average dinosaur, though, compared in size to a pony.

Speaking of elephants, an adult *Triceratops* weighed in at about the size of today's elephant. But instead of a trunk, it had three horns sticking up from its head like a mixed-up rhinoceros. Why three horns? Why not two? Or four? Or even five?

Five horns on its ugly face

Oops, some of them did. *Pentaceratops*, count 'em, had five horns on its ugly face: two cheek horns, two brow horns, and one lonely nasal horn. Knowing that radiation can affect bones more than any other body part, no wonder horn configurations can be so varied.

A few of those monsters could even fly. Some "dragons of the air" sported fur on their scales, while others—that couldn't fly—had feathers. But that's not so unusual. Ostriches and emus have feathers too—and they couldn't fly if their lives depended on it.

And some of those "flying reptiles" were huge. One of the largest, *Quetzacoatlus*, boasted a 63-foot wingspan, greater than the old twin-engine DC-3 airliner.

No ears emblazoned on their rears

Get the picture? With hundreds of genera of dinosaurs already identified, and more unearthed every day, you'd need a computer to keep track of them all. The point is, that even if they looked absurd, those ancient critters were put together with the same kinds of parts and in the same general order as the animals around us today. They didn't have arms growing out of their heads, or ears emblazoned on their rears, any more than we do. They weren't all that different. They just *looked* different.

The boom-boom-bellied Brontosaur had a tiny head about the size of a horse's. And yet, at 40,000 to 60,000 pounds, it weighed as much as 75 horses put together.

(Imagine putting a ping-pong ball where your head is.) How would you feed 75 horses with one tiny head, especially if that head contained only a handful of pencil-sized front teeth? With chompers like that, poor *Brontosaurus* probably couldn't even chew.

Some dinosaurs had gizzards

The New Mexican *Seismosaurus* had it even tougher. Measuring 140 to 150 feet long and weighing up to 180,000 pounds - as much as 225 horses - it too boasted a head the size of a horse's. (Imagine filling your hot tub with a spoon.)

How did poor Seismo survive?

It had a gizzard like modern-day birds, says Bakker. (Another reason to think birds are related to dinosaurs.) It swallowed its food whole, then swallowed rocks to grind the food. The rocks, called gastroliths, ranged in size from a peach pit to a baseball to a Texas grapefruit.

Once the rocks had served their purpose, the animals regurgitated them in one giant heave. "Dinosaur belches," the old-timers called them. That's why we find little piles of rocks all over dinosaur country. Utah State paleontologist David Gillette once counted 230 stomach stones in one Seismo skeleton.

But there's nothing unusual about gizzards. They're simply one more example of interchangeable parts.

Today's animals can be just as strange. Thumb through an old encyclopedia some rainy afternoon when you've nothing better to do, and take a gander at some of the peculiar-looking creatures inhabiting our planet today.

Why does the rhinoceros have a horn? Why do cows have two? Why does the javelina have short tusks? Why did mammoths have long tusks? What possible evolutionary purpose did the mammoth's giant tusks fulfill? Mammoth tusks 10 to 12 feet long were common, and 16-footers have been recorded. Sometimes the tusks grew into complete circles and toward each other until the tips actually crossed. But did those tusks serve a purpose? No, they were an oops, a burden, with which the mammoth learned to cope.

Did giraffes evolve their long necks so they could reach the leaves on tall trees? Or did they decide to eat those leaves *after* they were given long necks? Does the emu have three toes for a purpose? Then for what purpose does the ostrich have two? Did we develop our thumbs to play the guitar?

We showed up on this planet fully developed

No. We showed up on this planet fully developed, just the way we are. And so did they. Those who were lucky enough, or smart enough, learned to survive in *spite* of their handicaps, not *because* of them. Those who learned to live with what they were given—who learned to use their newfound colors, newfound shapes, new-found protuberances, and changed bones and teeth—survived. Those who said, "Half a wing be damned! If I can't fly, I'll walk," made it through. Those who didn't—died.

Did we evolve our seven neck vertebrae for a purpose? Then why does the giraffe also have seven? Why does the bat have the same number?

Why are the forelimbs of people, porpoises, bats, and horses, which look so different, and do such different things, built of the same bones?

It's like an assembly line. Add a part here, take one away there, paint this one blue, that one green, put a bigger horn on this one, no horn on that one, and five eyes on the next. No matter, they're all interchangeable parts. It may be a sedan, it may be a sports car, it may be a pick-up truck, but underneath, they're all built on the same basic frame.

Seals resemble dogs and whales resemble cattle. Elephants resemble sea cows and sea cows resemble modern-day conies (small rodent-like critters). If each of these animals "evolved," where are the intermediate forms?

They're not there.

No intermediate forms

"All paleontologists know that the fossil record contains precious little in the way of intermediate forms," said Stephen Jay Gould of Harvard. "Transitions between major groups are characteristically abrupt."

Same with humans.

As we dig through the fossil record, we push the arrival of humanids back further and further in time. Look at the Afar specimen (*Australopithecus afarensis*). Although the earliest known Afar specimen (our first upright-walking ancestor) is about 3.9 million years old, it's almost identical to Laetoli specimens. And yet, there's a million years between them.

A million years with no change

A million years with no change.
Where is evolution?
You see the same story everywhere you look. Almost every species on earth appeared suddenly, fully developed, in gigantic torrents of creativity, even the mighty Pterosaur. "Flying dragons seemed to burst into the world like Athena from the mind of Zeus, fully formed," says Robert Bakker. "Even the earliest pterodactyls already display fully developed wings and the specialized torso and hips so characteristic of the entire order." We find so many cases like this, says Bakker, that many scholars are persuaded that evolution doesn't work slowly at all. Sometimes evolution speeds up and "suddenly produces totally new adaptive configurations."

But how could Athena multiply? Asked Stephen Jay Gould. "With whom shall Athena born from Zeus's brow mate? All her relatives are members of another species."

It seems so simple, am I missing something big? Why couldn't she have been born a twin, or one of a litter? That's how a species must begin, as hopeful monsters born at the same time of the same womb.

Then the newcomers learn to cope. But they don't change their bodies to fit the circumstances, they adapt to the bodies they were given. It's not survival of the fittest;

it's arrival of the fittest. It's not survival of the fittest; it's survival of those who best learn to live with what they were given.

But what created those "hopeful monsters" in the first place? Geomagnetic reversals.

Geomagnetic reversals played important role

"The evidence is mounting," said James D. Hays of Lamont-Doherty Earth Observatory, "that Earth's magnetic field may have played an important role in development of life on this planet." (*Geol. Soc. Am. Bull. 82.*)

After each extinction, thousands of weird creatures must have appeared, as radiation worked its way up and down the ladder of genetic combinations. A few of those will-o'-the-wisp creations survived, but most were unworkable combinations. They died in the womb, or at birth, or had no mate, or were sterile, or were eaten, or who knows? The list of reasons for failure must have been endless.

Indeed, fossils of some dying families show an increase in size and an increase in the numbers of bizarre-looking forms right at extinction boundaries. An astonishing resumption of life took place at the start of the Tertiary, leading evolution into a kind of "experimentation run riot," says geophysicist Vincent Courtillot

Take the ammonite.

Near the end of its long reign, the ammonite experimented with a variety of shell shapes, said Professor Peter Ward of the University of Washington. Normally shaped like flat spirals, some ammonites changed to look like snails, others built long straight shells. Some grew into huge wagon wheels six feet in diameter or more; others shrank to the size of a silver dollar. It was a "mark of desperation," said Ward, "an attempt to find some way to win under the new rules."

Apes and humans branch apart

Those new rules were written by geomagnetic reversals.

Is it just a coincidence that apes and humans branched apart about five million years ago at the end-Miocene—at a magnetic reversal? Or that *Homo habilis* appeared about two million years ago (according to the Royal Tyrrell Museum in Drumheller, Alberta)—at a magnetic reversal? And an iridium spike? I think not.

Is it just a coincidence that *Australopithecus*, an upright-walking creature with a man-like jaw and an ape-like brain, went extinct about one million years ago (Royal Tyrrell Museum)—at a magnetic reversal? Or that *Homo erectus* (Peking man, or Java man) appeared about 780,000 years ago—at the Brunhes magnetic reversal? I think not.

Is it just a coincidence that *Homo sapiens neanderthalis* suddenly appeared about 115,000 years ago (Royal Tyrrell Museum)—at the Blake magnetic reversal? Or that the Neanderthal suddenly disappeared about 34,000 years ago (according to Niles Eldredge, Curator of the American Museum of Natural History)—at the Lake Mungo magnetic reversal? I think not.

No. Those sudden appearances and disappearances *had* to have had a common underlying cause. And that common cause, I submit, was radioactivity; radioactivity either allowed onto our planet, or actually created, by a reversing magnetic field.

Now let's explore how that radioactivity might have arrived on our planet.

Let's explore geomagnetic reversals.

> *Dinosaurs haven't gone extinct. You have a dinosaur bath in your backyard, roast dinosaur at Thanksgiving, and eat dinosaur nuggets at McDonald's.*
> —KEVIN PADIAN

7

RULER OF THE UNIVERSE

To see how powerful a magnetic reversal really can be, look at the sun. The sun, says Robert Noyes in his book *The Sun, Our Star*, is a giant unshielded nuclear reactor made of millions upon millions of never-ending nuclear explosions; nuclear explosions are the very engine that drives it.

But what causes those explosions in the first place? Magnetic forces are the cause, the very trigger, of those millions of explosions in the sun, says Noyes. Magnetic forces are the starter and the fuel that keep them going.

The explosions occur when hydrogen nuclei, or protons, in the sun fuse into helium nuclei. The protons whose fusion is required to start the reactions are mutually repelled, both by their electrical charge and by nuclear repulsive forces, says Noyes. Magnetic forces somehow defeat the repulsion.

Magnetic star

Our sun is a magnetic star. Like the earth, the sun has a north pole, a south pole, and an equator. And, also like the earth, the sun rotates.

It's hard to believe that something as big as the sun could rotate very fast.[1] But even with its huge diameter of 840,000 miles, the sun makes a complete revolution every twenty-seven days.

That speed, more than 4,000 miles an hour at the surface, creates millions of magnetic fields all twisted together like a huge ball of burning twine; millions of intertwined strands of intense, pulsating magnetic lines of force that heat the sun's corona to more than one million degrees.

One million degrees—all from magnetic forces!

---

[1] The sun is so huge "that a million Earths would fit comfortably inside," said Curt Suplee in "The Sun: Living With a Stormy Star."

Solar flares

Gigantic eruptions, called flares and prominences, continually burst from the sun, forming fiery loops many times the size of the earth and greatly increasing the already steady flow of particles in the solar winds.

There's so much energy in a flare that one flare—just one—could supply the energy needs of the entire United States for almost a million years. One flare can contain as much energy as two-and-a-half billion one-megaton hydrogen bombs.

Solar flares can throw arcs of blazing force across distances the width of Texas in seconds, and have been known to shoot 36,000 miles across the face of the sun in less than five minutes. At that speed, if they exploded on earth they'd traverse the United States east to west in 15 seconds, and encircle the globe in about three minutes.

Then the flares just sit there like massive bolts of lightning, crackling and popping in the sky. But unlike lightning, which disappears within a few seconds, these blazing arcs of force can survive for hours or days. Some solar flares last for weeks; others for months. Magnetic fields apparently support those loops of fire, said Noyes. They're "the invisible ropes that hold them up."

Northern Lights

Flares and prominences can shoot to staggering heights. During a flare on June 1, 1991, observers spied a bright jet of fire probing more than 400,000 miles above the sun, almost twice as far as it is from the earth to the moon. By the next day, said a story in the *Seattle Times*, a speeding cloud of particles, blasted free of the sun's gravity in the eruption, began squashing the earth's magnetic field and "northern lights lit up the skies as far south as Denver and Northern California."

Similarly, in October 2003, when the fourth most powerful flare ever recorded exploded out of the sun, the atmosphere was so electrically charged that the northern lights were seen as far south as the Mediterranean.

Auroras *always* occur immediately after solar flares, and are *always* associated with geomagnetic disturbances.

Did you know that?

Did you know that magnetic forces 93 million miles away (the distance from the earth to the sun) are the switch that turns on the northern lights?

Sunspots

Then come sunspots, those mysterious dark spots sometimes visible to the naked eye, which revolve east to west around the surface of the sun, the same direction that all planets move.

Ancient Chinese, Japanese, and Korean astronomers were aware of sunspots, and the Greek Theophratus, one of Aristotle's pupils, reported sunspots in 300 B.C. But sunspots didn't come under much scrutiny until 1610 when they were reported by Galileo.

No longer quite so mysterious, we now realize that sunspots are areas of extreme magnetic turbulence.

Sunspots are like oscillating magnetic dynamos. Their most predominant property is their intense magnetic field; it's the root cause of a sunspot's very existence.

The larger the sunspot, the stronger the magnetic field. The sun's normal magnetic field strength is about the same as the earth's, less than one gauss. As opposed to the sun itself, a sunspot's magnetic strength is immense, ranging from 500 to 4,000 gauss, far more powerful than the earth's.

Sunspots and magnetic reversals

Sunspots burst through the sun's surface every 11 years in gigantic nuclear explosions, says Noyes. And that's the point of this chapter. Those nuclear explosions are a direct result of a reversing magnetic field.

Ring all the bells and blow all the whistles. Those gigantic nuclear explosions in sunspots are the direct result of a reversing magnetic field.

At the beginning of each cycle, magnetic polarity on the sun reverses and magnetic north becomes magnetic south. No one knows why. Though the spots appear about every 11 years, there's a 22-year cycle involved. That's how long it takes for the polarity to reverse and then go back to its original state.

Maybe we don't understand why, but the simple fact still remains: magnetic reversals cause massive nuclear explosions on the sun. If magnetic reversals on the sun cause nuclear explosions, couldn't magnetic reversals on earth cause them too? I think they do.

Blasting into our skies at the speed of light, galactic-sized bursts of power reach earth almost immediately after a sunspot reversal, sometimes within nine minutes. Then they begin squeezing between the earth's protective magnetic lines of force.

The earth's magnetic lines of force hurtle out of the magnetic North Pole, take a long curving loop through space to completely surround the earth, then dive back in at the magnetic South Pole. Magnetic lines of force, whether belonging to the earth or to a common magnet, always form continuous closed loops.

Electrified particles rain to the earth

The solar wind (the electric current flowing from the sun) squeezes the earth's magnetic lines of force on the upwind side, and extends them on the down-wind side. Like a gigantic magnetic teardrop, the lines extend more than 40,000 miles into space on the daytime side of the earth (the side facing the sun) and more than 150 million miles on the nighttime side. Attracted to the earth's magnetic lines of force, then repelled when they get too close, the energized particles speeding in from the sun constantly corkscrew around the lines.

But the magnetic lines of force bunch together so closely at the North Pole - in an area known as the "cusp" - that the electrified particles have no room to continue their spiraling paths. This causes a few particles to leak out of the magnetosphere. Once released, like migrating geese headed north, they race helter-skelter into the earth's upper atmosphere where they precipitate toward the North and South Poles.

Upon reaching the upper atmosphere, the energized particles collide with its atoms and molecules, releasing radiation—much like that in neon lights—to create those undulating, waving, fluttering curtains of light that we call the aurora borealis.

The magnetosphere protects us

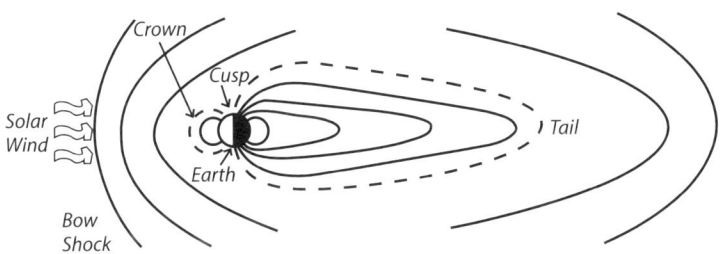

Auroras (from the Latin word for dawn) occur primarily in the far north or far south near the geomagnetic poles. (*Aurora borealis*, dawn of the north; *aurora australis*, dawn of the south.)

This process of escaping and colliding energized particles, this atomic bombardment, goes on all of the time, day and night.

Converting energy to matter

When an accelerated proton collides with an atom they merge, forcing the atom to split into two new nuclei. Usually called ionization, this barrage from the cosmos is also known as hydrogen impact ionization, or ionization by collision.

Whatever it's called, the collision unlocks the energy trapped in the charged particles and instantly converts it to matter. The same sort of thing happens in tests when an electric current is shot through gases, liquids, or solids.

But a flare greatly increases that stream of particles, squashing our magnetic field even more. The speeding particles, called solar cosmic rays, have energies as high as those in galactic cosmic rays.

As the field gets squashed, the lines push farther away from each other at the poles, creating "holes." This allows more particles into our atmosphere, ever closer to the equator, which in turn causes major increases in ionization as close as 47 miles above the ground, and causes the northern lights.

Though auroras can occur as high as 600 miles above the earth, and as low as 47, they usually occur about 60 miles up.

Isn't that kind of spooky, to realize that magnetic reversals 93 million miles away can trigger an atomic bombardment just 47 miles above your head?

That's during normal times.

Accelerating atomic particles to the speed of light

Some mechanism, so far not understood, preferentially accelerates these atomic particles to extremely high speeds. Accelerated to almost the speed of light (that should qualify as an "impossibly high rate of speed," don't you think?), some of the particles carry up to 10 billion electron volts. Somehow, scientists agree, the sun's reversing magnetic fields must play a major role.

Think about that for a minute!

Polarity reversals play a major role in generating up to 10-billion electron volts. Polarity reversals play a major role in accelerating atomic particles to the speed of light, the kind of speed that man-made nuclear particle accelerators can only dream about. And we don't think polarity reversals are important?

It's the same power that we use to create matter. Man-made particle accelerators, such as at the Stanford Linear Accelerator Center, create new matter almost routinely with headlong collisions between electrons and anti-electrons. Two miles long, the Stanford accelerator creates new electrons, *swarms* of new electrons, 40,000 times heavier than when they started. (It's a good thing electrons aren't people, isn't it? Stuff a two-hundred-pound human into an accelerator like that, and out of the other end would pop a mutant monster weighing eight million pounds.)

If a man-made accelerator only two miles long can create swarms of new electrons some 40,000 times heavier than when they started, imagine what a cosmic-sized accelerator could do. Ionization rates would increase by the billions.

And if a magnetic reversal 93 million miles away can trigger an atomic bombardment just 47 miles above our heads and create new matter, doesn't it seem possible that a magnetic reversal right here on earth, just zero miles away, could allow the solar wind to impact the earth even more, sending waves of explosions cascading around the globe and dumping untold amounts of radiation on our heads?

I think it does.

Carbon-14

Magnetic reversals on the sun do a lot more than paint pretty neon pictures in the sky. As sunspot activity waxes and wanes, so does the production of nitric oxide. So does the production of radioactive carbon-14. (Carbon-14 production increases during sunspot minimums.) In a direct example of energy turning to matter, part of the newly created carbon-14 enters the life-chain of both plants and animals. Trees grow taller and thicker, with their growth rings farther apart. The change in thickness occurs in exact synchronization with the sunspot cycle.

Sunspots Colder than the Sun

One surprising discovery is that sunspots are colder than the rest of the sun by several thousand degrees.

Colder. Not hotter.

Soon after a sunspot's magnetic field appears, the area gets cooler than the rest of the sun, and the flaring stops.

At the same time, the spectrum lines of many elements increase. Spectrum lines of carbon monoxide, calcium, silicon, fluorine hydrides, titanium and zirconium oxides, along with hydrides of nitrogen, carbon and magnesium, all show a much greater strength in sunspots. Such a strengthening, scientists believe, indicates an increased abundance of these elements.

This cooling has not been explained, but I think we're witnessing Einstein's theory at work ($E=mc^2$). Instead of burning matter to create energy, sunspots consume energy to create matter. One thing is certain, however; magnetic reversals in sunspots produce profound effects on the earth.

Imagine the consequences if a magnetic reversal should occur right here on earth, just zero miles away.

The style of evolution of brains-and much else-is not usually a matter of steady progression. Instead, the fossil record speaks of short periods of rapid and radical evolution, punctuating immense periods of time in which the sizes of brains hardly change at all.

— CARL SAGAN & ANN DRUYAN

8

....

PACEMAKER OF CREATION

....

What would happen if a magnetic reversal should occur right here on earth? For starters, there is credible evidence that radiogenic materials bombard our planet in sync with magnetic reversals. Second, there is credible evidence that magnetic reversals trigger ice ages. And third, there is credible evidence that all of those phenomena recur according to dependable, predictable, natural cycles.[1]

Radioactivity and our galactic orbit

Let's look at some of those cycles.

Peaks in the accumulation of radiogenic strontium correlate with our galactic orbit, as do peaks in lead, said Steiner and Grillmair of Canada's University of Alberta. Since uranium decays into lead, this means that accumulations of uranium correlate with our periodic trips around the galactic center.[1]

Major ice ages occur in sync with the same cycle. At least two Precambrian ice ages correlate with two of the six lead-isotope events in prehistoric Canada, said Steiner. All lead-isotope events, Steiner theorized, will eventually be tied to glaciation. (*Precambrian Res.*)

Our galactic orbit is not a perfect circle, Steiner explained, it's elliptical. Our distance from the galactic center varies from about 32,600 light-years when we're closest, to about 38,000 light-years when we're farthest away. We're approaching our closest position right now, and will reach it within eight million years, plus or minus four million. Since major glaciations occurred

---

[1] *Not by Fire but by Ice* (this author) presents solid evidence that geomagnetic reversals may indeed trigger ice ages.

[1] A full revolution around the galactic center—a cosmic year—takes about 282 million years.

about every 140 million years, said Steiner, "a causal relation appears plausible."

Radioactivity and our celestial orbit

Not only do ice ages and the accumulation of radioactive materials correlate with our galactic orbit, they also correlate with our celestial orbit.

Surprisingly, our orbit is not a circle, it's a stretched-out oval, so sometimes we're closer to the sun than others. And the shape of that oval constantly changes. Starting as an almost perfect circle, our orbit slowly stretches into an oval. Then, due to the gravitational pull of the planets, collapses back to nearly a perfect circle again.

Called orbital eccentricity, or orbital stretch, it takes about 100,000 years for our orbit to stretch from a circle to an ellipse and back to a circle again. Today our orbit is only slightly eccentric - about one percent - but it varies from a low near zero to about six percent. We're about 11 million miles farther from the sun when the stretch is greatest.

Ice ages and equinoctial precession

Ice ages also recur in sync with equinoctial precession. The seeds for that discovery were planted 150 years ago by James Croll, a self-educated Scotsman. Croll knew that the earth's axis of rotation is not perpendicular to the solar plane. Today we're tilted away from true north at $23\frac{1}{2}°$, but the tilt slowly increases to about $24\frac{1}{2}°$, then decreases to about 22°. The complete shift, back and forth, takes about 41,000 years.

Croll also knew that our axis of rotation wobbles like a top, tracing a clockwise circle around true north. Called axial precession, it takes about 25,800 years to complete the full circle.

Precession occurs, say scientists, because the sun and moon exert a gravitational pull on the earth's equatorial bulge. Rotating objects such as tops and gyroscopes also precess. So does Mars.

To understand this phenomenon, picture the globe spinning around a long stick (the axis of rotation). Tilted away from true north, the top of the stick traces a circle around true north, while the bottom makes an identical trip around Antarctica.

As our axis of rotation moves, it constantly points toward a different star, painting an imaginary circle on the heavens. The process of painting that circle on the celestial ceiling is called precession of the equinoxes.

Sir Isaac Newton solved yet another aspect of the riddle. The earth's orbit around the sun also revolves, said Newton. Our orbit revolves backward, or counter-clockwise. Precession of the equinoxes, the time it takes to paint that imaginary circle on the heavens, therefore takes about 23,000 years. It's like waiting for someone on a merry-go-round; you'll reach them sooner if you walk toward them.

Today when viewed from the northern hemisphere, the stars seem to rotate around Polaris, at the end of the handle of the Little Dipper. That's why it's called the Pole Star, because the North Pole points toward it.

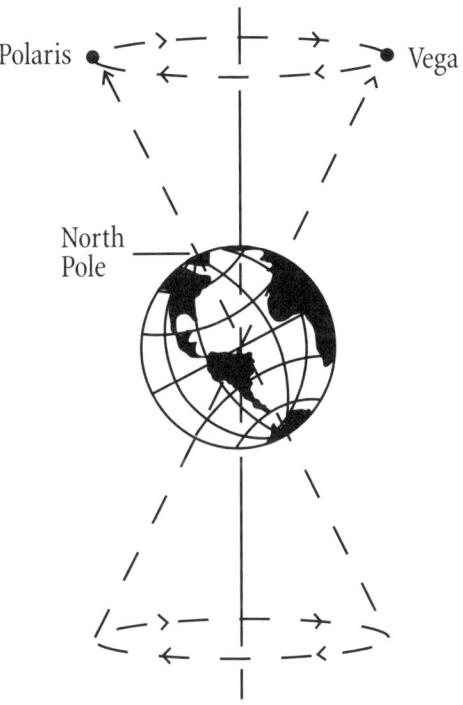

Precession of the Equinoxes

But in 2,000 B.C. the North Pole pointed toward a spot halfway between the Little Dipper and the Big Dipper. In 4,000 B.C. it pointed toward the end of the handle of the Big Dipper. Twelve thousand years from now it will point toward a different star, toward Vega, and in 23,000 years it will point toward Polaris again.

Scientists agree that equinoctial precession affects our magnetic field, but they don't know why. And many agree that magnetic-field intensity waxes and wanes in a cycle. "Dipole intensity fluctuations are periodic, with a period of about 10,000 years," said Allan Cox, in *Plate Tectonics and Geomagnetic Reversals*.

Geomagnetic field strength is falling

We appear to be nearing the end of just such a period right now. Geomagnetic field strength is falling dramatically. During the last 2,000 years, geomagnetic field strength has decreased by fifty percent, said Peter J. Smith in *The Earth*. Other scientists, such as Walter Elsasser of the University of Utah, peg the decline at sixty-five percent. One_half? Or two-thirds? Either way, it's a major decline.

Yet another concern is that the magnetic north pole is moving. The magnetic north pole has been drifting away from North America toward Siberia at such a clip, says Joseph Stoner, a paleomagnetist at Oregon State University, that Alaska might lose its spectacular Northern Lights in the next fifty years. The rate of movement has increased during the last century, Stoner added.

When will the reversal occur?

Just when our magnetic field will flip is unclear, but there's a growing body of evidence that it could happen sooner rather than later, said scientists at the Greenland Space Science Symposium in May 2007. "The best guess is that there are still several centuries to go," they said, pointing to the declining magnetic field strength.

But they don't know how long the reversal will last. It's not clear, the attendees said, how long the earth's protective magnetic shield will be down. "The record in the rocks is little help, since a geological eye-blink represents many human lifetimes." (*The Economist*, 12 May 07, p. 85-86)

Not only is geomagnetic field strength declining, the rate of decline is picking up. Five percent of the decrease has occurred during the last one hundred years alone. This suggests that we are experiencing either a fluctuation from the mean behavior, said geophysicists McFadden, Merrill, and McElhinny, "*or a precursor to a new reversal attempt.*" (Emphasis mine.)[1]

Upcoming magnetic reversal

A recent study published in *Nature Geoscience* found that the earth's magnetic field is weakening in several areas. "This may suggest the possibility of an upcoming

---

[1] McFadden and McElhinny are with the Australian Geological Survey. Merrill is at the University of Washington.

reversal of the geomagnetic field," said study co-author Mioara Mandea.[1]

Fluctuations in the magnetic field have occurred in far-flung regions of the earth, from Australasia to Southern Africa to the South Atlantic. "An oval-shaped area east of Brazil is significantly weaker than similar latitudes in other parts of the world," said Mandea, of the German Research Centre for Geosciences.

"What is so surprising is that rapid, almost sudden, changes take place in the Earth's magnetic field," said co-author Nils Olsen, a geophysicist at the Danish National Space Center.

The study also showed that the decline in magnetic field strength is opening the earth's upper atmosphere to intense charged particle radiation.

The reversal could come even sooner. "If present trends were to continue," said Harwood and Malin, of the Institute of Geological Sciences in Sussex, U.K., "the field would reverse in about 2230 A.D."

It could reverse tomorrow

Our magnetic field could reverse tomorrow. Here's why: As our axis of rotation makes its 23,000-year circle around the North Pole, magnetic intensity slowly increases, then decreases, every 11,500 years. Up and

---

[1] "Earths' Core, Magnetic Field Changing Fast," by Kimberly Johnson, *National Geographic News,* June 30, 2008.
http://news.nationalgeographic.com/news/pf/76158139.html

down it goes, turning in and out of sync with the solar system's magnetic field like a giant rheostat in the sky, just as a light bulb glows dimmer or brighter as you turn a dimmer switch.

Normally, magnetic intensity rises and falls gradually. And sometimes it rises and falls within the 11,500-year cycle itself. But toward the end of each cycle, magnetic intensity drops through the floor.

That's when the trouble begins.

In a study of lava flows at Steens Mountain, south-central Oregon (which erupted *during* a reversal, by the way), scientists Prévot, Mankinen, Coe and Grommé found that magnetic intensity had fallen to less than ten percent of today's intensity in less than one year.

Perhaps in less than two months.

During a follow-up study in 1989, Coe and Prévot found that the field had reversed at the rate of three degrees per day.

Perhaps in only three weeks

Not content with their earlier findings, Coe and his colleagues took another look. The earth's magnetic field had reversed at "the astonishingly rapid rate," their new study found, of six to eight degrees per day.[1] Not only did the field reverse, it fluctuated. Rapid fluctuations occurred

---

[1] The speed of the reversal is calculated by comparing magnetic field direction in different parts of the same lava flow. By estimating how long it took the center to cool as opposed to the faster-cooling edges, a time frame can be established.

many times during the reversal, said Coe. "Enhanced external [magnetic] field activity ... from the Sun might somehow cause the jumps." (Coe, Prévot, and Camps)

Such oscillations have also been noted in other locations. Valet, Laj, and Langereis found rapid geomagnetic-field fluctuations during a reversal in western Crete, while Michael Fuller found rapid fluctuations in the Tatoosh Intrusion on Washington's Mount Rainier. ("Ancient Magnetic Reversals: Clues to the Geodynamo," by Kenneth A. Hoffman)

Those kinds of magnetic field fluctuations, I submit, generate massive surges of electricity in and above the earth, causing gigantic explosions in the ground and in the atmosphere. Then radioactive materials begin raining from the sky.

How long the process takes, and whether it becomes a full or aborted reversal (an excursion), depends on where we're located in our stretched-out orbit around the sun, which in turn determines how strongly the sun's magnetic lines of force affect us, and therefore how destructive the reversal is.

Then snap! Since the earth can't turn upside down, our magnetic field reverses. Or, depending on where we are in our orbit, it continues fluctuating, generating ever more electricity until we move out of alignment once again. It happens twice per rotation, once on "this" side of the precessionary wobble and once again on the other, every 11,500 years. It's the same thing that happens to our solar system every 141 million years in its orbit around the galactic center, but on a much smaller scale.

It's a celestial game of orbital tag ... and we're "it."

Variations in the accumulation of beryllium-10 are probably caused by solar and geomagnetic modulation.
— G.M. Raisbeck

9

MAGNETIC REVERSAL CYCLE

You'll often hear that the most recent magnetic reversal occurred about 780,000 years ago at the Brunhes/Matuyama boundary. But that date is way off the mark. At least ten magnetic reversals and excursions, probably many more, have ravaged our planet during the past 780,000 years.

Laschamp magnetic reversal

In 1967, Norbert Bonhommet and J. Babkine discovered a geomagnetic reversal in lava flows at Laschamp and Olby, at Chaîne des Puys (chain of volcanoes) in central France. Our magnetic field reversed about twenty to thirty thousand years ago, they announced, and then remained reversed for about ten thousand years. They called it the Laschamp magnetic reversal. Is it just a coincidence, Bonhommet asked, that the return to normal polarity corresponded with the end of an ice age?

Though later research placed the Laschamp event at around 44,000 years ago, its discovery made us aware that other magnetic reversals or excursions might have occurred.

Gothenburg magnetic excursion

The most recent excursion, the Gothenburg magnetic excursion, occurred about 12,350 years ago (Mörner and Lanser). During that excursion, magnetic intensity fell dramatically, to about twenty percent of the Holocene average. (Mankinen and Wentworth, 2003) At the same time, magnetic inclination moved 180°. It also fluctuated, making wild swings of up 80° (Kopper).[1]

---

[1] Merrill and McElhinny placed the Gothenburg excursion at 8,000 to 14,000 years ago. Kenneth Creer placed it at 11,000 to 14,000 years ago. (*Earth Planet. Sci. Lett.*, 1976)

Mono Lake magnetic excursion

Another magnetic excursion, the Mono Lake excursion, occurred about 23,000 years ago (Kukla, Berger, Lotti, and Brown). During the Mono Lake event, magnetic intensity fell ten times faster than normal (Liddicoat and Coe).

Lake Mungo magnetic excursion

Before that came the Lake Mungo excursion of 33,500 years ago (Barbetti and Flude). And prior to that came the "real" Laschamp event of about 47,000 years ago, when magnetic intensity fell to less than 15% of today's (All magnetic reversals and excursions show major decreases in intensity. Roperch, Bonhommet and Levi, 1988)

See the cycle?

Those excursions struck like clockwork every 11,500 years, and they've been doing it for millions of years.

Geomagnetic reversals about 10,000 years apart have been found in the 65-million-year-old Deccan Traps, said geophysicist Vincent Courtillot. Indeed, 10,000-year hiatuses between lavas of opposite polarities are observed "frequently." (Watkins)

Other scientists agree. Magnetic intensity fluctuations of from two to 30,000 years' duration appear in the marine record as "tiny wiggles" and are therefore easy to overlook, said Steven Cande and Dennis Kent of Lamont-Doherty Earth Observatory (1992). We believe that this type of behavior of tiny wiggles, they said, "may have

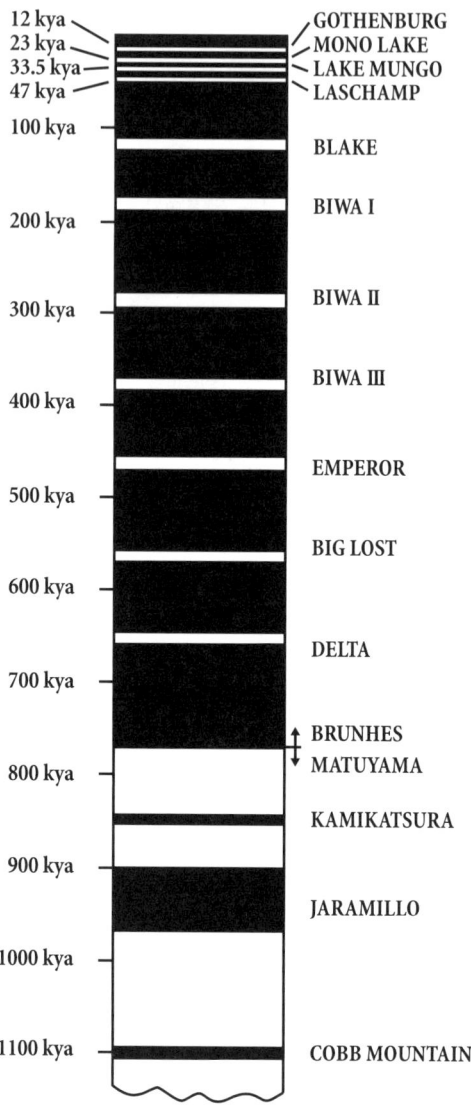

Magnetic Reversal/Excursion Chart
(After Champion *et al*)

characterized the geomagnetic dynamo throughout the Cenozoic [the last 65 million years]."
I think we'll eventually find millions of such "tiny wiggles" in the geologic record.

Geomagnetic reversals and ice ages

But here's the topper: Catastrophic cooling and rapid ice build-up accompanied many of those magnetic reversals and excursions. At least twelve ice ages can be correlated with magnetic reversals and excursions in the past two million years alone.

The Gothenburg magnetic excursion coincided with a period of short-term ice and snow, said Michael R. Rampino of NASA, as did the Lake Mungo excursion, when rapid cooling immediately followed a period of warmth.

The Mono Lake magnetic excursion coincided with glaciation; the Blake magnetic reversal at the end-Eemian coincided with glaciation, as did Biwa I, Biwa II, and Biwa III. (Champion, Lanphere, and Kuntz of the USGS discuss these magnetic reversals at length.)

Each of those catastrophic cooling episodes, said Rampino, "may have been triggered by a magnetic excursion. The Earth's magnetic field may be directly modulated by precession." (*Geology,* Dec 1979)

So there you have it.
Polarity reversals, equinoctial precession, and ice ages, all march to the same drummer. As do extinctions and

new species appearance. Toss in the specter of massive amounts of radioactivity falling on your head, and you've got the picture.

Look at the number of catastrophes that have befallen our planet in almost perfect sync with equinoctial precession during the last 115,000 years alone:

Catastrophes in sync with equinoctial precession
(kya = thousands of years)

115 kya - Blake magnetic reversal. Spikes in radioactive carbon-14 and strontium. Ice age begins abruptly following a period of warmth similar to today's. Sea levels surge 20 feet, then plunge at least 50 feet, in less than a century.

103 kya- Beryllium spike. Ice age ends. Major earthquake shoves Barbados and other islands higher.

91 kya - Beryllium spike. Ice age begins catastrophically from a climate even warmer than today's (Dansgaard & Duplessey). Heavy volcanic activity (Kennett & Huddleston).

80 kya - Ice age ends. Major earthquake shoves Barbados and other islands higher. Lake Missoula flood.

69 kya - Beryllium spike. Ice age begins abruptly. Rising reefs (Bloom). Yellowstone erupts (Christiansen).

58 kya - Beryllium spike. Mass extinction. Giant pigs, giant baboons, three-toed horses—all gone. Major volcanism (Dawson). Ice age ends.

43 kya - Laschamps magnetic excursion. Beryllium spike (three times normal). Carbon-14 spike (two times normal). - Firestone *et al*). Ice age begins abruptly. Rising reefs.

34 kya - Lake Mungo magnetic excursion. Beryllium spike. Carbon-14 spike (almost twice normal - Firestone *et al*). Short-term ice build-up, then ice age ends abruptly. Lake Missoula flood. Lake Bonneville flood. Intensive volcanism. Neanderthal disappears (Eldredge).

23 kya - Mono Lake magnetic excursion. Ice age begins abruptly. Major volcanism. Spikes in radioactive beryllium and carbon-14 (four to five times normal). Mass extinction. European forest elephant disappears. Mammoths clobbered.

11 kya - Gothenburg magnetic excursion. Mass extinction; 72% of large mammal species go extinct, whereas only 10% of small mammal species disappear. Spikes in radioactive carbon-14 (three

to four times normal), beryllium-10 (two to three times normal), iridium (two to three times normal - Firestone *et al*). Spikes in CO2 and many other elements. Rapid and severe ice build-up, then ice age ends in less than 20 years (Dansgaard) and today's warm period begins. Worldwide volcanism (Lamb). Nile River flood. Connecticut River flood. Lake Missoula flood. Lake Bonneville flood. Gulf of Mexico flood. St. Lawrence River flood. Worldwide tectonic uplift (Dawson). Creation of the Carolina Bays (more later).

And that brings us to today, frighteningly unprepared for the next beat of the magnetic-reversal cycle.

Now let me ask you: Knowing that so many violent catastrophes have descended on our planet in sync with geomagnetic reversals or excursions, and knowing that geomagnetic field strength has declined dramatically during the past 2,000 years, and knowing that the rate of decline has picked up, and knowing that the north magnetic pole is moving, and knowing that the rate of movement has increased, doesn't it make sense to consider what would happen to us during a magnetic reversal?

It is taken for granted that all the matter of the earth has been inherited from the time of its initial accretion. Each of these cognate assumptions is false; matter is created continuously and spontaneously at all levels.

—S. WARREN CAREY

10

·····

TUNGUSKA

·····

Dropped from a single B-29 bomber and set to detonate at 2,000 feet, the world's first atomic bomb exploded over the Japanese city of Hiroshima at 8:15 A.M., August 6, 1945. Pulverizing everything within reach, it destroyed almost two-thirds of the hapless city within seconds. Generating countless numbers of spontaneous fires and producing intense winds that fanned the flames, it killed 70,000 to 80,000 people almost instantly and left another 70,000 maimed and mangled.

We like to think of the Hiroshima explosion as the biggest explosion in history, but it wasn't. Thirty-seven years earlier, at 7:00 A.M., June 30, 1908, a far bigger blast shattered the stillness over Tunguska, central Siberia. The explosion, at about 26,000 feet, could be seen from hundreds of miles away. (That's like seeing Boston blow up from New York City, or Paris explode from London—in the daylight.)

1,000 Hiroshima bombs

Detonating with the force of 1,000 Hiroshima bombs, the force of the Tunguska explosion was so great that it caused a disturbance in the earth's magnetic field similar to those following nuclear explosions in the atmosphere, said Professor Clyde Cowan (with Atluri and Libby). Recorded on seismographs the world over, its shock waves shot twice around the globe. If that explosion had occurred in Chicago, the thunder would have been heard in Washington, D.C., North Dakota, Georgia, and Oklahoma, said Cowan, co-discoverer of the neutrino.

And it was *hot!* It charred the clothes of a farmer sitting on his porch 36 miles away, and burned a neighboring farmer's ears.

It looked like a pillar of fire and smoke, said observers, with a roiling, boiling, mushroom-shaped cloud rising sixteen miles into the cold Siberian sky. Then the cloud spread. Glowing particles hung in the skies over northern Europe for the next two nights as if reflecting from a luminescent cloud, while large amounts of its debris raced thousands of miles around the globe, reaching California in just two weeks. Skies glowed as far south as the Caucasus Mountains, glowed so brightly, said a local report, that "newspapers could be read at midnight without artificial light."[1] Readers wrote to *The Times* in London, remarking that its columns could be read outdoors at midnight," said Duncan Steel in a 2008 article in *Nature*. It took two months for the brightness to diminish and disappear.

When scientists reached the site of the explosion some twelve years later, they found three-foot-diameter trees flattened—snapped off like matchsticks or completely uprooted—over an area of 1,200 square miles. Such a blast would wipe out all of New York City and cause great destruction in neighboring Connecticut and New Jersey, said Baxter and Atkins in their book *The Fire Came By*. Amazingly, no crater has ever been found.

Exploding meteorite?

What caused the Tunguska explosion? Scientists at the Center for Relativity at the University of Texas chalk it up to "a black hole." Others, such as Cowan, Atluri and Libby, think it was an anti-matter nuclear explosion. Then there are those, such as Baxter and Atkins, who swear that it must have been an exploding flying saucer. Most researchers, however, lean toward an exploding meteorite.

---

[1] The Caucasus Mountains, near the Black Sea, are 2,400 miles from the blast site.

Tektites

But there's a problem with the meteorite theory. In all of the years since that early-morning explosion, no one has ever found a trace—none whatsoever—of a meteorite. Small, black, glassy spherules are the only things that have been found. Called tektites, the spherules are tiny black droplets of glass.

Sometimes shaped like teardrops, other times like baby barbells, tektites more commonly look like miniature marbles or glass buttons, flat on one side. Tektites can only be created, say scientists, in the extreme heat and pressure of an impact.

Interestingly, Tunguska's tektites contain high amounts of iridium. Iridium has also been found in Antarctic ice that formed at the same time as the Tunguska explosion, says Christopher Chyba of NASA.

Other places where tektites have been found seem to confirm the meteorite theory. Tektites have been found at the 15-million-year-old Rieskessel Crater in southern Germany, and again at the one-million-year-old Bosumtri Crater in Ghana.

We just can't get away from meteors, can we? That early-morning blast over Tunguska, and its tektites, sends us right back to the dinosaur extinction and to Mexico's Chicxulub Crater, site of the so-called meteor impact that killed the dinosaurs.

Chicxulub Crater

Similar black tektites, the same age as Chicxulub (tail of the devil), have been found at Beloc, Haiti. The Haiti tektites are impact splash droplets flung into the sky by the asteroidal collision, then blown from the Yucatán Peninsula to Haiti, say Alan Hildebrand and William Boynton of the University of Arizona.

The ejecta layer on the floor of the Caribbean is *packed* with tektites, *huge* tektites. Normally no bigger than a BB, Chicxulub's tektites are up to one-third of an inch long.

Tektites have also been found at other extinctions; at the dividing line between the Middle and Upper Jurassic, and again at the end-Eocene, when 13 billion tons of tektites were strewn from Georgia to the Indian Ocean.

Thirteen billion tons!

Scoop that many tektites into dump trucks, stack those trucks one on top of the other, and you'd have a rickety pile of dump trucks stretching 247,000 miles into the sky, farther than it is to the moon.

From one lonely asteroid?

More "proof" that tektites come from airborne intruders accumulates daily. Two scientists from the University of California, Phillipe Claeys and Stanley V. Margolis, along with Jean-Georges Casier of Belgium's Royal Institute of Natural Sciences, recently reported evidence that an asteroid caused the Devonian extinction of 367 million years ago.

Why an asteroid? Because of the microscopic glassy beads they found in end-Devonian black shale near the

Belgian town of Senzeilles. The beads, similar to the K-T tektites, have the varying shapes and a lack of bubbles that attest to their formation during an impact, said Casier.

The debate is over, say impact believers, the glassy spherules are impact products.

But if spherules are impact products, how do they explain the lack of an impact at Tunguska? No impact, just an explosion five miles high in the sky.

And no crater.

And no meteorite fragments.

Even if it was a meteor, why would it explode in mid-air? And with such force? No, I don't think those tektites came from a meteor. Nor do I think the K-T tektites came from a meteor. Nor the iridium.

Iridium at other extinctions

Take another look at that iridium. Iridium has been found, not only at the end-Cretaceous, but at many other mass extinctions; at the Callovian extinction, at the end-Devonian extinction (in Australia's Canning Basin), at the Tithonian extinction, at the Great Permian extinction, at the Precambrian, at the end-Eocene (when those 13 billion tons of tektites were strewn about), at the mid-Miocene, and again at the Late Pliocene extinction of two million years ago.

Iridium has also been discovered in places that impact proponents are hard-pressed to explain. Iridium has been found at the Ferguson Ranch in eastern Montana at the base of the lower Z coal, says paleobiologist Steven M. Stanley.[1]

Iridium at the base of a layer of coal

Another unexplained layer of iridium—at concentrations 250 times greater than normal—has been found in a swamp in the Raton Basin in northern New Mexico, at the base of a layer of coal. There is no known process, say perplexed scientists, that would create a single layer of iridium at the base of a coal bed.

So many iridium-laced boundary clays are associated with coal or carbon-like deposits that scientists wonder if there might be a connection. Articles by Block and Dams, and Kucha, and Wezel (in *Dynamics of Extinction,* D. Elliott, editor), point that way. The articles tell of finding excess iridium not only in coal, but also in kerogen in Permian limestones, and in mid-Cretaceous bituminous shales.

And in his book *Evolutionary Catastrophes* (p. 66), the French geophysicist Vincent Courtillot tells of the discovery of "a correlation between the distributions of iridium and carbon," and further, "that fossil bryozoans (generally marine animals that live in colonies) were encrusted with this carbon."

---

[1] Z coal, a kind of coal that formed right at the K-T boundary, has scientists scratching their heads. Why would coal form at the same time as an extinction? (More later.)

It's two anomalies in one. Why did so many carbon deposits form at the K-T boundary? And why do we find so much iridium in or near the same deposits? Some 200,000 to 300,000 metric tons of iridium were deposited on our planet at the K-T extinction, says Courtillot. Are we supposed to believe that all of that iridium came from one lonely meteor?

Excess osmium

All sorts of weird stuff rained out of those K-T skies. Excess osmium, for example, is found in end-Cretaceous clays in many locations, including Denmark and New Mexico. Excess osmium is even found in boundary clays at the Columbia River.

Where did that osmium come from?

Here we go again.

The osmium came from a meteor, say impact believers. Iridium and osmium, they insist, can only be produced in massive stars that have used up their supplies of hydrogen and helium. Once the hydrogen and helium are gone, the stars begin burning forbidden elements such as neon and carbon, thus starting fusion reactions.

Then the stars explode, sending huge chunks of themselves reeling into space, blasting iridium and osmium to every corner of the galaxy. Iridium, osmium, and all of the heavy elements must therefore have formed, say impact proponents, just before and during a brief, spectacular, and extremely hot supernova event. Iridium and osmium are literally, they say, the by-products of thermonuclear explosions. (Make a note of that in your memory bank. Iridium and osmium and all of the heavy elements are by-products of thermonuclear explosions.)

Therefore, the only way iridium and osmium could get to earth, impact believers insist, is if they were carried here by a meteor.

Strontium spike at K-T extinction

But what about all of the other metals in the boundary clays? Many boundary clays are rich in palladium, gold, nickel, and platinum. "Certainly more than would be expected from a meteoric source," says Charles Officer of Dartmouth College. There's a strontium spike too, says paleontologist Rob Thomas at West Montana College in Dillon, Montana. (Make another note. There's a strontium spike at the K-T extinction.)

And that's just the beginning.

Boundary clays at Caravaca, Spain, show spikes in cobalt, chromium, antimony, arsenic, and selenium, said Officer and Drake in *Science* (1985). Again, "it's more than you'd expect from a meteoric source." And then, of course, there's Wendy Wolbach's ubiquitous layer of soot right at the K-T boundary. (Wendy was the graduate student who found the worldwide layer of soot at the K-T boundary. See chapter 3.) Where did all of those elements come from?

They did come from the sky, I propose, but not from some rambling asteroid gone astray. All of those elements were deposited on this planet by massive explosions—

underground explosions, underwater explosions, and atmospheric explosions. Those elements were created, not on some untraceable exploding star a zillion miles away, but right here in our own skies above the dinosaurs' heads.
Just like carbon.

Carbon forms in the sky

That's right, just like carbon. I'm not guessing, nor proposing, nor theorizing here; it's a fact. Carbon does rain from the sky. We've known it for years.
Have you ever heard of radiocarbon dating?
Of course you have, unless you've been living in a cave. But you may not know how it works. Radiocarbon dating was developed in the late 1940s by Willard F. Libby at the University of Chicago after he discovered that carbon is constantly formed in the sky.

Carbon is continuously created in the atmosphere, day and night, rain or shine, 24 hours a day, by the interaction of cosmic rays and nitrogen atoms. Called nuclear electron capture (as mentioned earlier), it's a process wherein electrons are added to, or removed from, atoms or molecules that were previously neutral. Traveling at nearly the speed of light, the energetic particles in cosmic rays split the nuclei of atmospheric gases, creating the background radiation to which we are exposed every day. Add a neutron to almost any atom, said Libby, and it will become one unit heavier and frequently radioactive. Tiny amounts of radioactive carbon-14 are falling on our heads this very second.

Eventually, the newly created radiocarbon atoms get absorbed into the bodies of all living plants and animals, but only for as long as the plants and animals live. When the plants and animals die, the radiocarbon in the organic tissue immediately begins to decay. But the radiocarbon doesn't disappear. It changes into inert atoms of nitrogen at rates which, with the right equipment, can be measured.

Libby, a one-time member of the Manhattan Project, and former commissioner of the Atomic Energy Commission, reasoned that he could use the decay rate to find the time of death for any fossil. All he had to do was measure how many carbon atoms in the fossil were still radioactive. He was correct. Radiocarbon dating works remarkably well for any fossil less than 40,000 years old.

Now let me ask you. If we know that new matter—radioactive matter—is constantly formed in the atmosphere, isn't it possible, given the right circumstances, that the rate of accumulation could increase dramatically? Especially when we know that carbon is not the only element created in the sky?

Beryllium-10

Take beryllium-10. Radiogenic beryllium-10, a metallic element, is created—much like carbon—when cosmic rays smash into nitrogen and oxygen molecules over the north and south poles. The beryllium-10 then drifts to the earth where it becomes embedded in the polar ice sheets.

Another radioactive element, helium-3, is also formed in the sky.

Radiogenic hydrogen (tritium) is made the same way. So is potassium-40. Even iridium, scientists confess, constantly drifts from the sky.

It's a normal everyday occurrence. Newly created matter rains onto our heads all of the time. About 100,000 kilograms (220,000 pounds) of cosmic debris fall to the earth every day (Grieve and Sharpton).

Nearly 40,000 tons of microscopic particles rain down on the earth each year, agrees Gisel Winckler of Lamont Doherty Earth Observatory.

Forty thousand tons! Per year!

That's during a *normal* year.

What if something should happen to drive that creation process crazy? What if new matter were created in overwhelming excess during mass extinctions?

I think it is.

Atomic bombs release carbon-14

Oh, by the way, there's another way to make carbon-14. Atomic bombs release enormous amounts of radioactive carbon-14 into the air, said Sam Cohen in his 1983 book *The Truth about the Neutron Bomb*. Carbon-14 is one of the by-products of neutron bombs. For each 20,000 tons of TNT-equivalent, you get about two pounds of radioactive fallout. (Cohen should know what he's talking about: He invented the neutron bomb.)

That's why Cowan, Atluri, and Libby (the same Libby who developed radiocarbon dating) thought Tunguska was an anti-matter nuclear explosion. Knowing that carbon-14 is absorbed by plants, they looked at tree rings in the years immediately following the explosion. They found a spike in carbon-14 in 1909, the very next year. Perhaps Cowan and his colleagues were onto something. Perhaps it was an anti-matter nuclear explosion.

Come to think of it, with its luminescent skies and mushroom cloud, doesn't Tunguska seem suspiciously similar to a nuclear explosion?

It may well have been. According to A.V. Zolotov, a prominent soviet geophysicist who led several research expeditions to the site in the late 1950s, that blast over Tunguska was indeed a nuclear explosion.

Even those who think Tunguska was caused by an exploding flying saucer (Baxter and Atkins) think the explosion was "decidedly atomic."

Nuclear. Nuclear. Nuclear.

Uncanny, isn't it, how that word keeps cropping up? "Like a nuclear holocaust," said scientists of the K-T extinction. "Equivalent to the energy of one billion Hiroshima bombs," they reported. "Like a long nuclear winter," they said. "Like a Strange-love Ocean." "More damage than could have been caused by 10,000 times all of the world's atomic bombs." "The oceans were heated as though by a nuclear furnace."

Let's get off this meteor kick right now. If iridium and osmium and all of the heavy elements are by-products of thermonuclear explosions, who needs a meteor to explain them?

We had hot, violent nuclear explosions right here on our own planet, in the ground and in our skies, and Tunguska proves that it is possible.

Principal health hazard in radioactive fallout

Still want more proof?

Look at the strontium. With all of the headlines given to iridium, we've ignored one of the most important clues of them all. We've ignored the strontium.

What's so important about strontium?

Strontium is the principal health hazard in radioactive fallout. And we find it scattered all over the world right at the K-T boundary!

Not only at the end-Cretaceous, sudden increases in the amounts of radioactive strontium have also been found at other extinctions

Sudden increases in strontium occurred in the early Pennsylvanian, the early Triassic, and the early Tertiary, said Zell Peterman, in *Megacycles, Long Term Episodicity in Earth and Planetary History*. A marked shift in both carbon and strontium also occurred at the Great Permian extinction, said Douglas Erwin in *Nature*.

Call it mutation, call it creation, call it what you will, with so much radioactivity raining from the sky, the possibilities for new genetic combinations must be endless.

Really new trails are rarely blazed in the great academies. The confining walls of conformist dogma are too dominating. To think originally, you must go forth into the wilderness.
—S. WARREN CAREY

11

HUNDREDS OF THOUSANDS OF TUNGUSKAS

The change was electric, as if a switch had been thrown in the cosmos. Hundreds of thousands of ear-shattering, radiation-spewing, nuclear blasts exploded around the globe as a fiery inferno raced across the heavens.

Natural nuclear explosions.

Hundreds of thousands of Tunguska-sized pillars of fire and brimstone roared into the skies, unfolding layer upon layer of billowing mushroom clouds. Thick layers of deadly black fog encircled the globe turning day into night—and a nightmare—in an instant, a night that lasted for years for the planet, and forever for 70% of all living species. A dinosaur Sodom and Gomorrah.

If that blast over Tunguska was equal to 700 Hiroshima bombs, the K-T disaster was equal to millions more.

Underwater explosions

Explosions blasted upward from the bottoms of the seas throwing great swaths of seafloor violently into the sky along with billions of gallons of seawater and all of the animals in it, only to be pounded once again by other explosions reaching down from the clouds.

That's how huge "meteor" craters like Chicxulub were formed, from the inside out, blown outward by massive underground explosions.

It looked like the end of the world, and for millions of helpless creatures it was. Pulverized shells and shattered fish flew high into the sky, then fell back to the ground in a grisly rain of fragmented splintered bones.

That's how Denmark got its Fish clays—blown outward by massive underwater explosions.

Explosions on land sprayed millions of tons of dirt, trees, rocks, animals, dust, iridium, and anything else unlucky enough to be in the way, far into the sky, altering the landscape forever, leaving gaping holes in the ground to mark their violence.

Volcanoes—of a kind that we never knew existed.

Explosions in the sky

Explosions reached down from the sky, flattening trees, pulverizing and scattering rocks and boulders across the landscape, sparking forest fires around the planet, and fusing vast deposits of desert sand into the purest silica glass imaginable.

Yes, fusing desert sand into glass.

At least 1,400 tons of the purest natural silica glass ever found lies strewn across hundreds of square kilometers of the Libyan Desert. Some chunks of glass are larger than bowling balls and weigh as much as 57 pounds (26 kgs). Interestingly, the glass is very rich in iridium.

The sand from which the glass formed had melted "like sugar beneath a blowtorch," said researcher Giles Wright. Streaks in the samples indicate that the glass had sizzled at thousands of degrees and "flowed like a river" for more than a week.

Some scientists, of course, want to blame a meteor. But Dr. L. J. Spencer, member of a special expedition to the site in 1934, reported that no fragments of meteorites or indications of meteorite craters could be found. "It seemed easier to assume," said Spencer, keeper of Minerals at the British Museum, "that it had simply fallen from the sky."

More recently, when Vincenzo de Michele, keeper of minerals at Milan's Museum of Natural History, and

Romano Serra, an astrophysicist at the University of Bologna, made a thorough search of the site in 1996, they found the glass concentrated in two areas. One area is oval-shaped, while the other is a ring about 21 kilometers across and six kilometers wide. The area at the center of the ring is devoid of the glass, says de Michele.

The two Italian scientists theorize that heat from a huge explosion in the sky toasted the earth. "The ground could have responded in an elastic way to the blast wave and rebounded, leaving a ring and a central peak which later eroded," says de Michele.

Now, more than half-a-century later, the idea that explosions in the sky created an incredible amount of heat still appears to be the best explanation.

As you might surmise, I think such explosions remain the best explanation for the dinosaur's demise.

Einstein's theory applied

Explosions reached down from end-Cretaceous skies as fusion reactions cascaded around the globe. In a startling but logical application of Einstein's theory, powerful implosions sucked energy from the cosmos, then used it to create new matter. Carbon, iridium, antimony, chromium, strontium, arsenic, osmium, selenium, you name it, clogged the skies, obscuring the sun for months, while glassy black tektites tinkled to the ground.

Tektites—teardrops from the gods for all of the death and destruction.

How can we deny it? We have billions of tons of tektites to prove it. And we have spikes in all of those elements, precisely at the K-T boundary.
What we do not have ... is a meteor.

Creating new matter

Why go through all of these mental gymnastics trying to persuade ourselves that those elements came from some non-existent, non-findable meteor when we already know that nuclear explosions can, and do, create matter; when we already know that newly formed carbon and metallic ions are raining to the ground this very second?

That's how Mother Nature works. Sometimes it drizzles. Sometimes it pours. Sometimes violent thunderstorms rage through the skies. And sometimes, our everyday drizzle of carbon turns into a rampaging vicious storm. No longer measured by the ounce, it falls to the earth by the ton, flooding the entire globe.

Ten thousand times more carbon than normal

Boundary clays contain ten thousand times more carbon than normal. Did all of that carbon come from wildfires? No, it came from the sky, from a creation process gone horribly awry.

The most damning evidence comes from Wendy Wolbach herself. "The particle-size distribution of the soot is similar to that assumed for the smoke cloud of nuclear winter," said Wolbach. "No trace of meteoritic noble gases, and no meteoritic spinel were found in these carbon fractions."

Wendy was flirting with the truth and didn't know it.

Scientists have also found other anomalies in Wendy's worldwide layer of soot. Some of the K-T carbon formed at temperatures as high as 1,200°C and "is clustered in ways that could have come from hotter temperatures than would have occurred in a wildfire," said I. Gilmour in a 1992 article in *Science*.

What could be hotter than a wildfire?

Natural nuclear explosions.

Tremendous oil catastrophe

Forget those meteor myths. That carbon was created in our very own skies. Millions of barrels of soot and oil hemorrhaged from carbon-sodden clouds to be swept by torrential rains into every swamp and low-lying spot in the world, and to the bottom of every sea.

Then time worked its magic. During the ensuing millions of years the soot and oil turned to coal, kerogen, asphalt, black shale, lignite, and tar pits. Earth, coal bin of the gods.

Speculation? Yes. Fantasy? No.

There's a place in western Cuba—the Universidad Formation—that contains huge quantities of asphalt. In some areas entire banks of asphalt are found. In others, great masses of asphalt lumps lie around. The asphalt was deposited on the seafloor about 50 million years ago, said

Björn Kurtén in his book *How to Deep-Freeze a Mammoth.*
It is "clear evidence that a tremendous oil catastrophe occurred in the far distant past," said Kurtén, a professor of paleontology at Finland's University of Helsinki since 1955.[1]
Naturally occurring asphalt is also found at Pitch Lake in Trinidad, the only known asphalt lake on our planet.
Natural oil spills are a fact.

Carbon from the sky

So is carbon from the sky.
Your trusted teachers were wrong. Wrong, when they told you that coal was formed in ancient swamps by the build-up of layer upon layer of decaying plants. Wrong, when they told you that the layers had grown so thick and heavy that they had compressed themselves into coal.
The fact is that no one really understands how coal is made. That "coal-was-made-in-a-swamp" business is pure speculation and guesswork. "The chemistry of coal is still not well understood," said Carl Sagan in *Parade*.
If coal is made of crushed vegetation, why does it so often contain no plant remains? For the most part, baffled scientists admit, there are no macroscopically recognizable plant remains in coal. Coal in the Red Deer River Valley of Alberta, Canada, for example, "contains very

---

[1] Is it just a coincidence that a geomagnetic reversal occurred at about the same time as the Cuban oil catastrophe? (Kent and Gradstein) I think not.

little conspicuous plant matter," said D. W. Gibson in *Geological Survey of Canada*. Plant remains are found above the coal, plant remains are found below the coal, but almost no plant remains are found in the coal itself.

Standing Trees

None, that is, except for countless numbers of well-preserved trees and their roots along with standing trees extending upward through the coal seams.
Same in the Ukraine.
The coal of the Donetz Basin of the Ukraine contains fossilized tree trunks that extend through a coal seam from the carbonate rock below the coal to the layer above the coal. These fossils are coalified where they are within the coal seam and are not coalified where they are in the carbonate," said Thomas Gold in his 1987 book *Power from the Earth*. (This is not a new discovery. Charles Lyell mentioned trees standing in coal seams more than 160 years ago.)

If coal is made of decomposed plants, why didn't the trees and their roots decompose along with the other so-called plant materials?
Because there *weren't* any other plant materials! That carbon was deposited on *top* of the trees. That carbon came from the sky!
How else would we explain the locations of so many of the world's coal seams? Admittedly, most coal seams are layered between sedimentary strata, but many are not. Layers of coal have been discovered buried between

layers of volcaniclave, most notably in southwestern Greenland (Pederson and Lam). Other coal deposits, with no indication of sedimentary compression, have been found in New Brunswick, Canada, where the coal fills an almost vertical crack that extends downward through many horizontal sedimentary layers (Hitchcock).

Brown coals and lignites have always refused to fit into the swamp-believer's theories.[1] If coal was formed by all of the weight above it, how do we explain the brown coals of Moscow? Moscow coals are so close to the surface, scientists grudgingly admit, that there isn't enough material above them to have compressed them into coal.

So how do they explain it?

---

[1] Lignite, also known as brown coal, is the lowest rank of coal. It has a high moisture content and very high ash content compared to bituminous coal.

Bituminous coal, a relatively hard coal, contains a tar-like substance called bitumen. Of higher quality than lignite but lower than anthracite, it is usually black but sometimes dark brown, often with well-defined bands of bright and dull material

Anthracite is a hard, compact coal with a high luster. It has the highest carbon count and contains the fewest impurities of all coals. It is also referred to as blue coal, hard coal, stone coal, blind coal (in Scotland), Kilkenney coal (in Ireland), crow coal, and black diamond. The imperfect anthracite of north Devon and North Cornwall in England is known as culm.

Brown coals and lignites must have been created through some special form of creation, they mumble, through some special form of genesis.

You bet they were formed by "a special form of genesis." Natural nuclear explosions!

Radioactivity and oil

Same with oil. No one really understands how oil is made either. "It is suspected," say scientists, "that natural radioactivity may be instrumental in oil formation."

Let me repeat: Natural radioactivity may be instrumental in oil formation.

Same with natural gas. The formation of natural gas, say scientists, may be due to radioactivity. Hey, maybe this natural nuclear explosion theory isn't so crazy after all.

Natural nuclear explosions

Multi-layered, multi-colored fireballs bombarded the earth, from ground zero to 600 miles in the sky.

Imagine the mayhem. Stampeding in panic, vast herds of dinosaurs thundered across the plains. Trampling each other as they strove to escape, they dove for cover in any rocky crevasse, burrow, or cave they could find. Plunging into rivers and lakes, over cliffs, under trees, under rocks, they tried to escape the crescendo of sound, a symphony of sorrow and death from the heavens.

The ones that couldn't fit under the rocks, the ones weighing more than fifty-five pounds, were mostly instantly killed. "Anything that survived must have been in a burrow or under water—or very lucky," said Jan Smit of the Free University in Amsterdam.

But were they really so "lucky?"

Wounded and beaten, their world destroyed, their food supplies gone or contaminated, most were rendered infertile. The "lucky" ones lived to see their offspring born dead. Or deformed. Or mutated.

Brand new creations.

Creation

Creation—that's the operative word here.

New types of animals soon appeared that had never before existed on the face of the earth, to slowly re-colonize a depleted planet.

You see the same scenario after every mass killing in history. New forms of life immediately followed each extinction. Plants and trees were flattened and incinerated. But their roots and seeds, exposed to massive doses of radiation, soon burst forth in a twisted diversity of newly created genetic combinations.

And why not? We engineer new plants in our labs every day. Shouldn't nature do it even better?

It was not "a scenario analogous to that of a neutron bomb." It *was* a neutron bomb. "Equivalent to the energy released by at least one billion Hiroshima bombs," said Alvarez.

No, Mr. Alvarez, it was not "equivalent" to one billion bombs, it *was* one billion bombs.

"As much energy as would be released by five billion Nagasaki-sized bombs ... about 26 bombs per square mile," said Allaby and Lovelock in *The Great Extinction*.

No, gentlemen, it was not as much as 26 bombs per square mile, it *was* 26 bombs per square mile.

Five thousand? Five million? Or 26 bombs per square mile? It doesn't matter. What we need to know is what set those bombs off in the first place. How do you "light" a nuclear explosion?

Electromagnetic forces.

Almost every man-made fusion reaction on earth, say physicists, is triggered by electromagnetic forces. And that's what we're talking about here; electromagnetic forces run amuck.

Electricity ruled the world

Racing through the water at the speed of light, deadly currents of electricity transformed every ocean in the world into a vast conductor of death. Instant ionization flashed through the seas, sending billions of molecules into frenzied frantic collisions as electrostatic polarities reversed, breaking the ionic bond.

Millions of lightning strokes danced through the heavens, slapping the world in the face. Popping, sizzling, crackling, sparking, like welding torches in the hands of the gods, they destroyed one form of life, then used the pieces to create yet another.

Creation in the midst of destruction. Do we cry for the dead, or do we cheer? Without those deaths, mutations, and new creations, we wouldn't exist.

Finally, the bombing stopped. But the ground, littered with broken trees, twisted bodies, and shattered lives, had only begun to die. The worst was yet to come from that deadly radiation suspended in the clouds, deadly radiation that would rain on the earth for months if not years.

Now we know why so many dinosaurs have been found twisted and contorted as if poisoned by some unknown toxin—radioactive poisoning.

Dinosaurs and uranium

Countless numbers of dinosaurs, far more than can be dismissed as mere coincidence, have been unearthed next to unexplained deposits of uranium. Where did that uranium come from, if not from nuclear reactions?

And where does the uranium in coal come from? Yes, coal. Some kinds of coal, mystified scientists admit, contain huge amounts of uranium. Brown coals and lignites contain up to 0.1 percent uranium. "By far the greatest human-made contribution to radioactive pollution is not leakage from the wastes and cooling water of nuclear power plants but uranium-rich plumes from the smokestacks of coal-fired power stations," said the late Thomas Gold, a respected astronomer and professor emeritus at Cornell University.

Or consider the association between uranium, helium, and oil. "Helium derives mainly from ongoing radioactive decay of uranium and thorium," said Gold, which might explain why helium concentrations in oil-bearing areas are often a hundred times greater than in neighboring ground.

A hundred times greater!

Where did that uranium come from? From a nuclear firestorm in the sky, from hundreds of thousands of Tunguskas.

Our world—our expanding globe—was not created during some unknowable, unexplainable "Big Bang"; it was, and still is, being created by accretion, one tiny speck at a time, by millions of "Little Bangs," all triggered by geomagnetic reversals.

In the beginning, was the rotating electromagnetic field.

The quantity of black coal and petroleum (and especially its natural gas component, methane) are far greater than could be explained by any theory that depends on buried biological debris.

—Thomas Gold

12

....

CARBON RAIN

....

"I know a world midway in size between the Moon and Mars," said Carl Sagan, "where the upper air is crackling with electricity; where the perpetual brown overcast is tinged an odd burnt orange; and where the stuff of life falls out of the skies like manna from heaven onto the unknown surface below."

And what is that "stuff of life" that Sagan is talking about? That "manna from heaven"?

Hydrocarbons and nitriles constantly fall from Titan's skies, said Sagan. Titan - the big moon of Saturn - is socked in as a haze of organic solids formed high in its skies slowly fall and accumulate on its surface. Oceans of water are impossible on Titan (it's too cold), but "vast oceans of liquid hydrocarbons are expected."

Created, in other words.

"It's enough to make a Texas oil man drool," exclaimed an article in the *Seattle Times* (21 Mar 1995). New images from the Hubble space telescope show that Titan may have lakes of oil as big as all five Great Lakes put together.

Rivers of oil

It may be oil, or it may be methane.

Photographs taken by the European Space Agency's Huygens probe, which landed on Saturn's largest moon on January 14, 2005, show images of streams, springs and deltas that look eerily similar to river networks on earth, except that these networks were carved into the landscape by rivers of oil or liquid methane. Other images from the Cassini mission show hydrocarbon lakes, replete with shorelines, bays and channels.[1] One lake, as big as North America's Lake Ontario, has been dubbed Ontario Lacus.

---

[1] For more on the Casini mission see
http://space.com/scienceastronomy/050917_titan_shore.html

We estimate that Titan "contains more hydrocarbon liquid than the entire known oil and gas reserves on Earth," says Ralph Lorenz of Johns Hopkins University's Applied Physics Laboratory. "Titan sports a complete hydrological cycle, one where it rains methane," said an article in *Sky and Telescope.* (April 2005) The methane "evaporates, condenses, forms clouds, and rains back down onto Titan." Other hydrocarbon byproducts form a photo-chemical smog in Titan's atmosphere.[1]

Same on Jupiter.

Our experiments in ionizing a reduced atmosphere show that "it rains crude oil on Jupiter," said Willard Libby, in his 1969 talk "Space Science" (the same Libby who discovered radiocarbon dating).

---

[1] **Bitumen raining from the sky**

Suddenly, the old Mexican myths about bitumen raining from the sky (*The Manuscript Quiché*, Brasseur, *Histoire des nations civilisées du Mexique*, I., 55), or the old Syrian tales about oil raining from the sky (*Ras-Shamra [Ugarit]*, C. H. Gordon, *The Loves and Wars of Baal and Anat,* 1943), don't seem quite so mythical.

Nor do the Midrashim texts that speak of naphtha (petroleum) falling from the sky (*Midrash Tanhuma, Midrash Psikta Raboti, and Midrash Wa-Yosha*).

Immanuel Velikovsky told of these myths in his book *Worlds in Collision,* pp. 69-71 and 149.

Uranus and Neptune also have large admixtures of carbon in their atmospheres, said the late Thomas Gold in his 2001 book *The Deep Hot Biosphere: The Myth of fossil fuels.* "It is now generally agreed that there is a profuse supply of hydrocarbons on many other bodies of the solar system, where no origin from surface biology can be suggested," said Gold. "Carbon is the fourth most abundant element in the universe and also in our solar system. I am sure," Gold added, "that there were no big stagnant swamps on Titan."

Why not here?

It seems such a simple question.

Why not here?

If carbon can form in Titan's hazy skies, if crude oil can rain out of Jupiter's skies, then why not here?

Indeed, it looks as if one of the requirements will have been met. According to Donald E. Scott in his book *The Electric Sky*, Titan has "no discernable magnetic field."

Unfortunately, during a magnetic reversal, neither will the earth. As I mentioned earlier, all magnetic reversals and excursions show major decreases in intensity.

Prior to previous magnetic reversals our magnetic field strength dropped to about 15 percent of normal, and then suddenly went through zero and flipped. That will be the time of carbon creation, because that's when we also will have no discernable magnetic field.

Let's face it, the entire world is made of carbon; the entire universe. "We're made out of star dust," says Swiss physicist Dr. Walt de Heer of the Institut de Physique Experimentale. "A large proportion of interstellar dust is carbon. It's the stuff of the universe."

People are made of carbon, animals are made of carbon, trees are made of carbon, plants are made of carbon, 18% of all living things on our planet are made of carbon.

Even comets.

Comet dust contains bits of organic material similar to tar or soot, says Scott Sundfold of NASA's Ames Research Center in California.

Where did that carbon come from? It was *created* high in our skies; then fell to the earth to be incorporated into our bodies, into the bodies of all living things. When we die, when they die, we contribute our bodies to an expanding earth.

Ashes to ashes and dust to dust, we bestow our bodies to the earth. And when the soil turns to rock, we become fossils. More than 20% of all sedimentary rocks are made of carbon. Travertine is a calcium carbonate. And limestones and dolomites, at more than 50% carbonate, are also made of carbon, as are marble, sandstone, and chalk. Even tektites have a high content of carbon.

Black shales laced with carbon

Even black shales. K-T black shales contain ten times more carbon than all of the world's known coal and petroleum reserves combined, said W. F. Ryan of

Lamont-Doherty Earth Observatory. The Late Mesozoic may have been "the major carboniferous period of the earth since the Precambrian."

Many scientists think the black shales were created by anoxic waters. Organic matter in the water, they theorize, deprived of oxygen and therefore unoxidized by bacteria, imparted the black color of carbon to the muds.

The *color* of carbon? Just the *color*?

No way! Black shales are *carbon-rich*! They're *filled* with carbon! The alum shales at Kinnekulle in Sweden, for example, contain so much carbon that local lime burners once burned the shales outright as fuel.

And legend has it that American Indians and pioneers once used chunks of black shale to light their campfires. A visitor center in Parachute, Colorado, tells of a homesteader who built a black shale fireplace in his new home, only to watch his house burn to the ground after lighting a fire for his housewarming party.

Where did the carbon in those black shales come from? From the sky.

Carbon from the sky

Look at the black shales covering the huge coral-stromatoporoid reefs of Western Australia. When the corals died at the end-Devonian their death was sudden and widespread, said Steven M. Stanley in his book *Extinction*. Western Canada's reefs died abruptly, too. But were anoxic waters the culprit? Stanley is dubious. "It is far more plausible," said Stanley, "that the black shales were super-imposed on the reefs."

Right on, Mr. Stanley! And how do you superimpose carbon on a reef? It had to have come from the sky.

And it must have come fast.

Why fast?

Because of the squashed skeletons. Most graptolite skeletons beneath the black shale are flattened, said David Raup and Steven Stanley in their book *Principles of Paleontology*.[1]

Now, you don't flatten something by gently enfolding it with sediments. You flatten something when you hammer on it, or dump something on it fast and furiously.

Does it seem as if I'm splitting hairs here? Then why are graptolites *not* flattened when they're found in limestone?

Black shales are highly radioactive

Black shales contain one other deadly secret, though, far more persuasive than all of the other arguments put together. Black shales, baffled scientists admit, are 550 percent more radioactive than other sedimentary rocks.

Some black shales (also called kerogens because of their tar-like molecules) are worth mining for the uranium alone. In Sweden, oil shale is the raw material upon which the entire uranium industry is based. Or look at the highly

---

[1] Graptolites were small floating animals that went extinct about 325 million years ago.

radioactive black shales in the Antrim-Chattanooga-Woodford area of the United States.

Or consider the black shales of the Cambrian (the Kohm Shales). With a uranium content ranging from fifty to one hundred parts per million, the Kohm shales contain up to 200 grams (7.05 ounces) of uranium per ton.

"There does seem to be a correlation between layers of carbon and uranium," said Professor S. Warren Carey in his 1989 book *Theories of the Earth and Universe.*

How did those black shales become radioactive?

Natural nuclear reactions

Millions of years ago a water-cooled natural nuclear reactor simmered and spluttered in the rocks of Gabon, West Africa, for nigh on one million years, said Howard Brabyn, in his 1985 paper *A Precambrian Nuclear Reactor*. This startling conclusion, said Brabyn, was reached by nuclear physicists at the French Commissariat a l'Energie Atomique (CEA).

Uranium deposits at Oklo, near Gabon, contain less than the normal amount of uranium-235.[1]

But why?

---

[1] For more information about Gabon, read *A Natural Fission Reactor* by George Cowan, one-time head of the Nuclear Chemistry Division at Los Alamos Scientific Lab.

Also read Rao, K.R., 2002, "Nuclear Reactor at the Core of the Earth! - A solution to the riddles of relative abundances of helium isotopes and geomagnetic field variability, *Current Science*, Vol. 82, No 2, 25 Jan 2002.)

A chain reaction must have produced plutonium, the French scientists decided. The plutonium then decayed back into uranium, just as occurs in man-made nuclear reactions.

The uranium filled huge depressions in the area about 1.7 billion years ago. Then powerful tectonic upheavals ripped through the basin, changing its shape and reconcentrating the ores. It all adds up, the specialists agreed, to ideal conditions for a spontaneous nuclear reaction.

It seemed impossible.

But when the skeptical French physicists discovered four extremely rare elements in Oklo—cerium, neodymium, samarium, and europium—with compositions "hitherto found only as a result of man-made nuclear reactions," they could deny the facts no longer. A natural nuclear reaction *must* have occurred, they announced. Nature, not man, had constructed the world's first nuclear-fission reactor. Eventually, six such reactor zones were identified in Oklo.

Natural nuclear reactions have also occurred in other parts of our planet. Take the Carswell structure in northern Saskatchewan. Located about 70 miles south of Uranium City, the Carswell structure contains at least five important uranium deposits.

Surrounded by a circular sedimentary basin about 25 miles across, and flanked by precipitous cliffs from 50 to 210 feet tall, the Carswell structure is sometimes mistakenly identified as a meteor crater.

But it's not.

The Carswell structure was formed by a crypto-explosion (an explosion of unknown origin), said K. L. Currie in 1969. Support for Currie's astounding proposal came from M. Pagel *et al* of the Centre de Recherches sur la Geologie de l'Uranium. Shock features in the underlying rocks, said Pagel, "suggest that the event originated from below, not above."

Germany's Ries Crater also originated from below.

So did the Decaturville structure in Missouri.

The Steen River circular structure in Alberta, Canada, formed the same way, as a result of a "violent internal explosion," said James Lyons and Charles Officer of Dartmouth College.

So did Vredefort Dome in South Africa.

So did the Sudbury structure in Ontario. Sudbury's shock deformation features "indicate an endogenic [explosive or tectonic] origin," said Neville Carter at Texas A&M.

End-Cretaceous craters formed the same way. The Manson structure in particular, said Carter, "is clearly not an impact crater." It was caused by an internal explosion event and has uplifted about 3,000 meters.

An internal explosion event … of a kind that we never knew existed.

Natural nuclear explosions.

Nobody has yet synthesized crude oil or coal in the lab from a beaker of algae or ferns.
— THOMAS GOLD

13

....

DINOSAUR TOMBSTONES

....

The most telling proof that natural nuclear explosions have occurred lies in the diamonds. Nanometer-sized diamonds (one-billionth of a meter) have been found in K-T boundary clays all over western North America, said U.K.

scientist Iain Gilmour in a 1992 article in *Science*. Boundary clays at Knudsens' farm in Alberta's Red Deer Valley contain nanometer-sized diamonds, as do end-Cretaceous clays at Berwind Canyon, New Mexico, and at Brownie Butte, Montana.

Miniature dinosaur tombstones

Where did those miniature dinosaur tombstones come from?

Since Gilmour and his colleagues knew that primitive chondrites and urelites (ancient stony meteorites) are often "littered with diamonds," they first assumed that the K-T diamonds were relict interstellar diamond grains carried to earth by a meteor or asteroid.[1]

But the U.K. scientists soon changed their minds.

Carbon and nitrogen compositions of the K-T diamonds differed from meteoritic diamonds, they found. The end-Cretaceous diamonds were also larger and spanned a greater size range than those found in meteorites. "Meteorites are not likely sources of the K-T diamonds," said Gilmour. "They do not match known meteorite diamonds even remotely."

---

[1] Meteoritic diamonds were probably formed in cool areas of the primitive solar nebula by electrical discharges between dust aggregates, experts believe, then cooled rapidly to their glassy state. *Lightning, in other words, zapping back and forth between clouds of dust, can create diamonds.*

Diamonds produced in the sky

Were the diamonds blown from the ground by the asteroidal impact? Not likely. Isotopic compositions of the K-T diamonds do not match those of any major carbon reservoir on earth, said Gilmour, nor of any other forms of reduced carbon at the end-Cretaceous. "The K-T diamonds were most likely produced during the impact of the asteroid with Earth," said Gilmour, "or in a plasma resulting from the associated fireball."

Produced! Those diamonds were produced—meaning *created!*—right here on earth, or in the sky above it, precisely at the K-T boundary!

But how?

"Diamonds produced by the explosive detonation of TNT," said Gilmour, "have morphologies, grain size, and high N contents similar to those of the K-T diamonds."

Diamonds in the sky 12,000 years ago

If that were the only time that diamonds had rained from the sky, it might not be such a concern. But microscopic diamonds have also been identified from about 12,000 years ago.

The evidence for these nanodiamonds, presented by nuclear physicist Dr. Richard Firestone and geologist Dr. Allen West at a meeting of the American Geophysical Union (AGU) in 2007, came from more than two-dozen sites across North America.

The smaller diamonds, the "size of cold viruses," would have lingered in the atmosphere for weeks or months, said West. "The larger diamonds were visible to the naked eye and dropped like hailstones."[1]

"Diamonds drizzled down by the tons," said West. The skies must have rained precious stone and metals for several months. "These would have been like ten thousand Tunguskas going off at once."

The minuscule diamonds, along with high levels of iridium, were found embedded in a black mat of carbon spheres and carbon glass deposited at the start of a major cooling period about 12,000 years ago known as the Younger Dryas.

Creation of the carbon spheres would have required tremendous temperatures and pressure, temperatures of at least 4,000 C, said professor James Kennett, who also spoke at the meeting. This points to the explosion of an extraterrestrial object up to five kilometers across, said Kennett, of the University of California.

The explosion and ensuing wildfires could have caused the extinction of several large American mammals at the end of the last ice age, said Kennett. "All the elephants, including the mastodon and the mammoth, all the ground sloths, including the giant ground sloth—which, when standing on its hind legs, would have been as big as a mammoth" disappeared. "This explosion could also have had an enormous impact on human populations," said Kennett.

---

[1] http://www.livescience.com/strangenews/080707-canada-diamonds.html

Nuclear irradiation

The black mat in which the nanodiamonds were discovered was highly radioactive. "Our research indicates that the entire Great Lakes region (and beyond) was subjected to particle bombardment and a catastrophic nuclear irradiation that produced secondary thermal neutrons from cosmic ray interactions," the AGU scientists said.

"No skeleton of extinct megafauna has ever been found in or above the black mat, only below it, and no Clovis artifact has ever been found in or above the mat either," says Dr. Vance Haynes, professor emeritus at the University of Arizona. "The American horse, dire wolf, saber-toothed tiger, American camel, mammoth, and mastodon, all of them disappeared in an instant before the black mat formed. When we dig up their bones today, the black mat covers them like a blanket."

At the Murray Springs Paleo-Indian archeology site near Sierra Vista, Arizona, the black mat measures from a few inches thick to as much as a foot. We've found the black mat at dozens of sites around the San Pedro Valley covering hundreds of square miles, Haynes said. "Always, we find the Clovis artifacts and extinct animal bones right under it."

Uneven and lumpy, the black mat covered everything. When Haynes and his crew dug through the mat into one of the larger bumps, they uncovered a large mammoth that they dubbed 'Big Eloise.' "The mat was draped over her bones like a thick blanket," said Haynes, "staining her bones almost black and conforming to her skeleton as if it was shrink-wrapped." The black mat also conformed perfectly to hundreds of mammoth footprints in the area.

Iron balls from the sky

In addition to the black mat and its nanodiamonds, the researchers found tiny rusty iron balls deeply embedded in mammoth tusks, bison horns, arrowheads and chert. (Chert is a compact rock consisting essentially of microcrystalline quartz).

In their 2006 book *The Cycle of Cosmic Catastrophes,* Firestone and West, along with Simon Warwick-Smith, a field exploration geologist from Australia, tell of looking at these tiny iron balls under a microscope. Normally ten times narrower than a human hair, the iron balls appeared to have radiated outward from an overhead explosion. "Literally trillions" of those tiny lethal ball bearings had indiscriminately bombarded ice-age bison, mammoths, and people at hypervelocity, said Firestone.

As with the black mat, the Clovis-era chert was highly radioactive, containing significantly elevated levels of uranium and plutonium. Such high plutonium levels are usually found only around atomic-bomb test sites or nuclear reactors, said Firestone, of the Lawrence Berkeley National Laboratory.

Radioactivity 2,000 times normal

When archaeologist Dr. William Topping, who collaborated with Firestone, saw the high amounts of radioactivity in the chert, he decided to test the soil at a Clovis-era site in Michigan. Radioactivity levels in the Michigan soil measured 2,000 times higher than normal.

Extremely high levels of radioactivity, up to 1,600 times higher than normal, have also been discovered in Clovis-era soils in New Mexico, "rich enough for a world-class uranium mine," said Firestone.

Dr. Topping remembered that the Tunguska explosion had "peppered the ground with millions of tiny particles," some of which were found embedded in tree trunks facing the blast. Had this been a Tunguska-like event? he asked.

Moving 4,000 miles per hour

Tests conducted by the National Superconducting Cyclotron Laboratory at Michigan State University show that the tiny iron particles might have been moving as fast as 4,000 miles per hour, said Firestone, "almost certainly enough to knock a mammoth trunk-over-heels backward."

The tiny missiles had also come in hot. The researchers noticed brown rings around the entry points in the tusks and horns that indicated charring.

Evidence now suggests that these red-hot particles, traveling at jet speed, crashed into Siberia and most of the northern part of the globe, "maybe all of the Northern Hemisphere, including Europe," Firestone said.

Buckyballs

Nanodiamonds are not the only form of carbon found at the mammoth extinction. Firestone also describes the discovery of fullerenes, tiny soccer ball-like carbon cages. Also called buckyballs—after architect Buckminster Fuller, the inventor of the geodesic dome—fullerenes are created only occasionally "from intense natural forces such as *lightning strikes* (italics added), and rarely, inside some lava flows."

The buckyballs and radioactivity mark the occasion so well, said Firestone, that "armed with only a magnet and a geiger counter we found the magnetic particles in the well-dated Clovis layer all over North America where no one had looked before."

The Carolina Bays

A host of unusual geological features—elliptical sand-rimmed depressions collectively known as the Carolina Bays—appear to have formed at about the same time, said Firestone.

Varying in size from one to several thousand acres, and measuring from 50 meters (164 feet) to 11 kilometers (6.8 miles) across, as many as two-and-a-half million of these elliptic depressions scar the landscape from Florida to New Jersey to Texas.

In Maryland, the bays are called Maryland basins. In Mississippi and Alabama they're called Grady ponds. In Kansas and Nebraska they're called Rainwater basins. In Texas they're called Salinas (because they often contain salty water). The bays, aligned with one another with their long axes generally pointing north, all appear to have formed at the same time from the same cause.

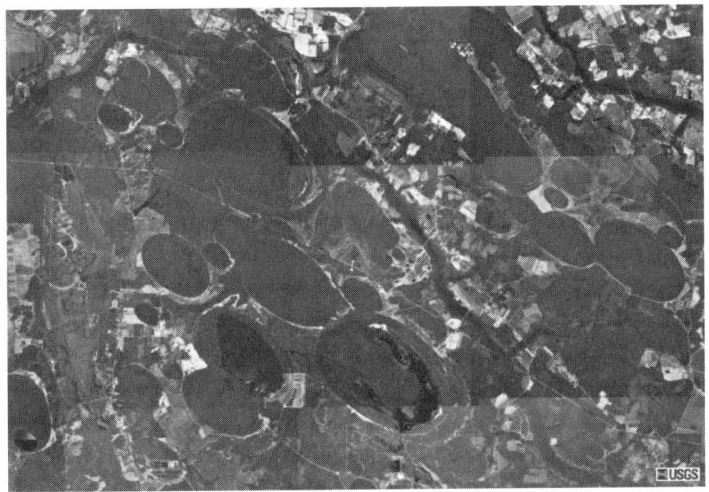

Carolina Bays   Image courtesy of U.S. Geological Survey

If you go to the USGS Terra Server website[1] and peruse the North Carolina/South Carolina region, you'll be astounded at how many of these giant paw prints you can find. The image above is of an area about 20 km by 15 km (12 miles by 9 miles) located approximately 30 km (18 miles) southeast of Fayetteville, North Carolina. There are more than twenty such depressions in this one tiny corner of the world. You can find similar views – more than you'd believe - on Google Earth. Literally hundreds of such bays surround Lumberton, North Carolina.

---

[1] http://nationalmap.gov/gio/viewonline.html

The bays all have raised rims, and frequently intersect other bays. More than 50,000 overlapping bays—some of them larger than nearby cities—have been identified on the United States Atlantic coast alone. Most bays are very shallow, and for some reason, the sandy rims are thickest on their southeastern ends.

The shallowness of the bays—fifty feet at most, but usually not more than twenty—makes it likely that they were not formed by meteoric impacts. Besides, almost no meteorites have ever been found in the Carolinas.

Yet another peculiarity is that the sandy rims are white. This, in spite of the fact that white sand is rare in the Carolinas; it's usually tan or reddish.

What could have turned the sand white? One explanation is chemical action, said Firestone. "The other is very high temperature, usually above 1,500 degrees F, which burns off the reddish iron impurities from the surface of the quartz."

All this, from some unknown, unproven, unexplainable extraterrestrial object? Or a supernova, as Firestone and his colleagues would have us believe?

I don't buy it.

Firestone and his colleagues have made some startling and important discoveries, but I don't agree with their interpretation of those discoveries.

A rogue asteroid didn't bag the short-faced bear. A meandering meteor didn't doom the dinosaurs. A star-crazed comet didn't snuff the sabre-toothed cat. Nor will some sneaky extraterrestrial object or supernova cause our own untimely demise.

A magnetic excursion

But what about a magnetic excursion?
Now that makes more sense.
Let's look at the dates.

According to Nils-Axel Mörner and Johan Lanser, the Gothenburg magnetic excursion occurred about 12,000 years ago. It doesn't take a huge stretch of the imagination to see that there might be a link between that excursion and the huge amounts of carbon, iridium, nanodiamonds and radioactivity that suddenly appeared on our planet.

Especially when we remember the Lake Mungo magnetic excursion of 32,500 years ago with its increased levels of radioactivity and tiny iron missiles, and when we remember the Laschamp magnetic excursion of around 43,000 years ago (Firestone et al.) with its increased levels of radioactivity.

Do we really believe that a supernova just happened to come along in sync with each of those magnetic excursions? Do we really believe that an unidentified "extraterrestrial object" just happened to come along every 11,500 years in sync with equinoctial precession?

A more logical explanation is that each of those magnetic excursions triggered millions of explosions above the mammoths' heads.

THEMIS

The idea of explosions in the sky recently got a big boost from NASA.

NASA launched a fleet of spacecraft in early 1997 to study eruptions of Northern Lights called substorms. One unexpected result of the mission, dubbed THEMIS,[1] was that the satellites observed small explosions in the earth's magnetic bow shock where the solar wind first feels the effects of the earth's magnetic field.

"Sometimes a burst of electrical current within the solar wind will hit the bow shock and—Bang! We get an explosion." said David Sibeck, project scientist for the mission at NASA's Goddard Space Flight Center in Greenbelt, Maryland.

Magnetic ropes

The satellites also confirmed the existence of giant magnetic ropes - twisted bundles of magnetic fields – connecting the earth's upper atmosphere directly to the sun. "We believe that solar wind particles flow in along these ropes, providing energy for geomagnetic storms and auroras," said Sibeck.

During a substorm over Alaska and Canada in 2007, the THEMIS mission spotted auroras surging westward twice as fast as anyone thought possible, crossing 15 degrees of longitude in less than one minute. "The storm traversed an

---

[1] THEMIS: Time History of Events and Macroscale Interactions during substorms

entire polar time zone, or 400 miles, in 60 seconds flat," said Vassilis Angelopoulos, the mission's principal investigator at the University of California, Los Angeles.

Angelopoulos estimated the total energy of the two-hour event at five hundred thousand billion Joules, equivalent to the energy of a magnitude 5.5 earthquake.[1]

The first magnetic rope that THEMIS encountered was very large, about as wide as the earth, and was located approximately 40,000 miles (70,000 km) above the earth in an area called the magnetopause.[2] The rope formed and unraveled in just a few minutes, providing a brief but significant conduit for solar wind.[3]

If such explosions can happen in today's world, just imagine how close to earth those explosions could occur during a magnetic reversal when we've almost completely – perhaps completely - lost our magnetic shielding.

---

[1] When describing the explosions, NASA didn't define the word "small." But if the energy of the two-hour event was equivalent to the energy released by a magnitude 5.5 earthquake, the explosions must have been at least as large as that early-morning blast over Tunkuska.

[2] The magnetopause is where the solar wind and earth's magnetic field meet and push against one another.

[3] www.nasa.gov/mission_pages/themis/auroras/northern_lights.html

Giant paw prints

Traveling southward at thousands of miles per hour, the explosions gouged those giant oval paw prints into the ground known as the Carolina Bays, dumped millions of tons of carbon onto our planet, and vaporized who knows how many millions of human beings, only to be followed by months - if not years - of radioactive rain.

That's why the bays are elliptical, because the shock waves were moving so fast.[1] That's what created the thicker rims on the southeastern ends of the bays; the shock waves shoved the sand in that direction. That's what burned the impurities from the sand. That's what propelled those red-hot iron missiles into the mammoth's tusks. And that's where those diamonds came from; explosions in the sky.

What are diamonds made of?

Just what are diamonds made of, anyway? They're made of graphite (pure carbon) subjected to extremely high temperatures and pressure.

The hardest naturally occurring substance known to man, a diamond is a close-knit, dense, strongly bonded structure vastly different from graphite. Ranging from colorless to white to black, diamonds can be transparent, translucent, opaque, yellow, brown, and rarely, can be blue, red, pink, green, or even orange.

---

[1] Please recall that the pure natural silica glass in the Libyan Desert was also found in elliptical-shaped areas.

Not content with nature's meager supplies, humans have learned to make diamonds, too. James Hannay, a Scottish chemist, produced the first manufactured diamonds in 1880 when he heated a strange concoction of paraffin, bone oil, and lithium in sealed wrought-iron tubes.

It would take another 75 years before an economically feasible way was found to make diamonds. Heating graphite to more than 3,000°C and subjecting it to pressures of 1,500,000 pounds per square inch, the General Electric Research Laboratory in Schenectady, New York, successfully produced man-made diamonds in December 1954.

Diamonds produced with explosives

One of the most important things about human-made diamonds, for this story at least, is how they're made.

Diamonds are produced with explosives.

Industrial diamonds, the kind we use for polishing and grinding, are produced routinely and in commercial quantities (more than 600 tons annually) by the controlled action of explosives on carbon.

Let me be very clear about this; I am not guessing. Explosions can—and do—create diamonds. And that, I submit, is what created the K-T diamonds; explosions in our own skies, right above the dinosaurs' heads.

Just as those microscopic diamonds of 12,000 years ago mark the mammoths' graves, just as those minuscule diamonds of 65 million years ago mark the dinosaurs' graves, let's make sure that yet more flying diamonds don't become our own miniature tombstones.

Now, if diamonds, the hardest naturally occurring substance known to man, can form in the sky, wouldn't it make sense that softer forms of carbon, such as Wendy Wolbach's layer of soot, could be created even more easily, and in far greater abundance?

That's why we find so many dinosaurs buried in coal.

Yes, dinosaurs buried in coal.

Dinosaurs buried in coal

While developing a new gallery in a Belgian coal mine near the French border in Bernissart, miners uncovered a remarkable concentration of Iguanodon skeletons at 322 meters (1,046 feet) below ground, said Edwin H. Colbert in his book *The Great Dinosaur Hunters and their Discoveries*. Driving another tunnel some 10 stories deeper, the miners found still more bones. The boneyard was obviously "of gigantic proportions."

The skeletons weren't indiscriminate piles of bones either, said Colbert, Curator of Vertebrates at the Museum of Northern Arizona. For the most part, the skeletons were remarkably well articulated. Instead of being scattered about as bones often are, they were still grouped together as if the animals had been buried whole.

Unusual? Not really. Dinosaurs and carbon go hand-in-hand. At the Geraldine Bone Bed in Texas, for instance, fossil skeletons are mingled with segments of fossil logs and dark carbon-rich stains that permeate the entire mass of sediment.

What unknown force could have solidified that carbon into coal so quickly?

Heat.

Apply enough heat by cracking the hydrocarbon chains, and coal will turn to oil and gas. Apply enough heat, and oil shale will turn to oil. Called hydrogenation, that's how Germany made 75% of its synthetic oil from 1941 to 1944. Apply enough heat, shock, and pressure to carbon monoxide and methane, and they'll disassociate into graphite and diamonds, said W. J. Nellis of the University of California's Lawrence Livermore National Laboratory.

If the small amounts of heat that humans have learned to generate can accomplish all of those things, imagine how quickly a few billion volts of electricity could turn a pile of carbon into coal.

There is simply no way around it; there is a close association between the dinosaur extinction and carbon.

On display at the Royal Tyrrell Museum of Palaeontology in Drumheller, Alberta, is an 18-inch-thick layer of black coal-bearing rock that formed immediately above the K-T boundary. Below the coal, dinosaur bones abound. Above the coal, are none. That could explain why so many dinosaur bones are black.

Z coal

Same with eastern Montana's Z coal. The Z coal marks the boundary between the Hell Creek Formation and the Tullock Member of the Fort Union Formation. Dinosaur bones are found throughout the Hell Creek Formation, said geologists Baadsgaard and Lerbekmo of Canada's University of Alberta. Above the formation, are none.

Doesn't a connection seem likely? Especially when there's so much iridium beneath the coal? Iridium values at the base of the coal are 200 times greater than normal, said Bruce Bohor with the USGS in Denver. (Not only with coal, "there is also a strong association of iridium with oil wells," said Cornell University's Thomas Gold.)

Even the vegetation changed, precisely at that layer of coal. A major change in plant life, widely documented in North Dakota, Montana, and Wyoming, occurred right at the lowest persistent lignite bed, say geologists.

What's going on here? Soot, diamonds, coal, iridium, black shale, and unexplained deposits of uranium, all buried beside—or actually encasing—the dinosaurs.

Where did that uranium come from? Where does *any* radioactivity come from for that matter? There must have been—there *had* to have been—millions of carbon-forming, radioactivity-spewing explosions all over the globe, right at the K-T boundary.

With so much radioactivity falling to the ground, is it any wonder that entirely new genetic combinations appear in the geologic record immediately following mass extinctions? Call it nature, or call it the hand of God, the idea of slow, stately evolution is not the only answer.

The biological molecules in oil show that the oil is contaminated by living creatures, not that the oil was created by living creatures.

—THOMAS GOLD

14

WHAT NOW?

Where does this leave us? Shouldn't we be searching for ways to avoid the dinosaurs' fate? Let's begin with the fact that Charles Darwin was wrong, that evolution does take leaps, creating new plants and animals never before seen on the face of the earth. Then let's move on to the fact that many of those leaps can be correlated with magnetic reversals and extinctions, and that many radioactive materials and hydrocarbons appeared on our planet at the same time.

And let me ask you again: "If hydrocarbons can rain out of Titan's skies, if crude oil can rain out of Jupiter's skies, then why not here?"

It did happen here.

The proof lies all around us.

The proof lies all around us

The proof lies in the unexplained deposits of uranium, strontium, iridium, beryllium 10, and carbon-14.

The proof lies in the unexplained layers of soot, oil, coal, black shales and diamonds.

The proof lies in the dinosaur skeletons entirely encased in coal. The proof lies in the ancient tree trunks standing upright in the coal seams.

The proof lies hidden in the giant oil fields of Alaska and Saudi Arabia. The proof lies interspersed with the tar sands of Alberta. The proof lies buried in the coal mines of Kentucky. The proof lies submerged beneath the deep blue waters of the Gulf of Mexico.

The proof sits waiting in your driveway, in the gas tank of the vehicle that you'll be driving to work tomorrow morning.

Will barrels of oil rain on my head?

Am I worried about barrels of oil raining on my head? No. As near as I can tell, major accumulations of oil are extremely rare occurrences tied to our solar system's periodic trips through the galactic arms, and to its periodic up-and-down bobs through the galactic plane. I don't think we'll see a repeat performance of that magnitude for another eight million years, a prospect that does not overly concern me.

With that said, I do believe that the possibility of a magnetic reversal or excursion dumping soot and miniature diamonds on our heads—along with increased amounts of carbon-14, strontium, beryllium 10, and other radioactive materials—is all too real. One way or another—blame a meteor, blame an asteroid, or blame a magnetic reversal—scientists agree that those elements did in fact come from the sky.

Whatever the cause—call it the hand of God or call it nature—the fact remains that geomagnetic field strength has declined by two-thirds in the past 2,000 years.

And the fact remains that the rate of decline is picking up, meaning that we could be much closer to the next magnetic reversal than we realize.

Remember, magnetic field strength doesn't need to drop all the way to zero before it reverses. During previous reversals, when magnetic field strength dropped to about fifteen percent of normal—then wham!—our magnetic field suddenly flipped.

Unfortunately, it's not as if we can just sit here and ignore the decline. The earth won't allow such passivity, because as I said in *Not by Fire but by Ice*, I think our magnetic field holds tectonic forces in check.

As magnetic field strength declines, I expect tectonic activity—earthquakes, tsunami, above-water volcanism, and underwater volcanism—to increase dramatically. This, along with ever more extreme swings in weather, will force us to pay attention.

Indeed, I think those extremes have already begun. Volcanic activity is now the worst in 500 years. Worldwide flood activity is the worst since before Christopher Columbus. In the U.K., flood activity is the worst in 1,000 years. In Poland, it's the worst in 3,000 years.

It's not all doom and gloom

Be that as it may, it's not all doom and gloom. We've been so busy trying to comprehend what killed the dinosaurs, and what caused so many of the other extinctions, that we've almost ignored what traits the survivors may have shared.

Maybe they were lucky, or maybe they were chosen, but so many animals made it across the K-T boundary, and so many animals survived the other extinctions, that they must have left clues in the geologic record as to how they did it.

The dinosaurs—and millions of other animals—may have gone extinct at the end-Cretaceous, but eighty-six percent of all freshwater animals survived. What was it about fresh water—as opposed to salt water—that helped protect them?

Most animal species where the adults weighed more than 55 pounds went extinct at the end-Cretaceous. What was it about their smaller size that saved them?

The Neanderthals may have gone extinct at the Lake Mungo magnetic reversal, but we—modern humans, who lived in Europe at the same time—survived. (How else would those old Mexican myths of bitumen raining from the sky have been passed from generation to generation if someone hadn't survived?)

The European forest elephant may have gone extinct at the Mono Lake magnetic reversal of 23,000 years ago, but our ancestors survived.

One million mammoths and mastodons may have died at the Gothenburg magnetic excursion of 11,500 years ago, but our ancestors survived. What did they do that helped them make it through?

Was it because humans lived in covered dwellings of some kind? Might our homes therefore be our salvation? How about our cars?

What can we do to prepare?

Something blasted those two-and-a-half million paw prints—the Carolina Bays—into the ground. Those depressions are there. They're measurable. They're quantifiable. And they're impossible to deny. *Something* has been attacking our planet every 11,500 years.

I don't like the thought of hundreds of thousands of Tunguskas exploding over my head any more than you do, but we can't just ignore it. We *must* recognize the danger

Just as we once threw our resources into landing on the moon, we must mount a concerted effort to discover ways to protect ourselves from a geomagnetic reversal.

Fall-out shelters?

What can we do about the ever-increasing amounts of radioactive materials that will almost certainly fall from the sky? Would adding lead to our roofing materials help protect us? Would individual protective suits help? Would we be safer if we stayed indoors? Would underground bunkers help? Should we activate old Cold War fallout shelters as planned in Huntsville, Alabama?[1]

Should we monitor the explosions taking place in the outskirts of the earth's magnetic field? Perhaps that would give us enough warning to head for our bomb shelters.

What can we do to protect our food supplies? Would it be enough to simply cook our food a little longer? Would a one-year supply of food be enough?

---

[1] Huntsville is now creating the most ambitious fallout-shelter plan in the United States, featuring the Three Caves Quarry, an abandoned limestone mine big enough for 20,000 people to take cover underground.
(See http:/abcnews.go.com/US/wireStory?id=3660316)

What about water? Should we build protective covers over our water reservoirs? Could we install special filters on our faucets? Would it help to boil the water? (Probably not. According to nuclear physicist Dr. Richard Firestone, boiling will not remove the radioactivity and other toxins.) Is there a chemical that we could put in the water to counteract the radioactivity?

Let's take a lesson from nature and do the same thing that the low-metabolism animals did so many years ago—pull into our shells and wait it out. If we can discover how they accomplished that feat, I think we'll have a good shot at survival.

Our ancestors made it through

Our ancestors didn't know about polarity reversals. Our ancestors didn't know about equinoctial precession. Our ancestors didn't know about strontium, or uranium, or any other kind of radioactivity. And yet, they made it through. If they were able to survive the holocaust with their limited knowledge, then I see no reason why we can't do the same. Provided, of course, that we pay attention.

The answers are out there. We need to find them.

Oh, and a little bit of luck wouldn't hurt.

Good luck.

BIBLIOGRAPHY

Allaby, Michael, and Lovelock, James, 1983, *The Great Extinction*.
Alvarez, L. W., Alvarez, W., Asaro, F. and Michel, H. V., 1980, "Extraterrestrial cause for the Cretaceous-Tertiary Extinction," *Science*, Vol. **208**, p. 1095-1108.
Alvarez, Walter, 1986, "Toward a Theory of Impact Crises," *Eos*, Vol. **67**, p. 649, 653-655, 658, 2 Sep 1986.
Alvarez, Walter, Alvarez, L. W., Asaro, F. and Michel, H. V., 1984, "The End of the Cretaceous: Sharp Boundary or Gradual Transition?" *Science*, Vol. **223**, p. 1183-1186, 16 Mar 1984.
Alvarez, Walter, and Asaro, Frank, 1990, "An Extraterrestrial Impact," *Scientific American*, p. 78-84, Oct 1990.
Alvarez, Walter, and Muller, Richard A., 1984, "Evidence from crater ages for periodic impacts on the Earth," *Nature*, Vol. **308**, p. 718-720, 19 Apr 1984.

Baadsgaard, H. and Lerbekmo, J. F., 1980, "A Rb/Sr age for the Cretaceous-Tertiary boundary (Z coal), Hell Creek, Montana," *Canadian Journal of Earth Sciences*, Vol. **17**, p. 671-673.
Baadsgaard, H. and Lerbekmo, J. F., 1983, "Rb-Sr and U-Pb dating of bentonites," *Canadian Journal of Earth Sciences*, Vol. **20**, p. 1282-1290.
Bakker, Robert T., 1986, *The Dinosaur Heresies*.
Barbetti, M. and Flude, K., 1979, "Geomagnetic variation during the late Pleistocene period and changes in the radiocarbon time scale," *Nature*, Vol. **279**, p. 202-205, 17 May 1979.
Barbetti, M., 1979, "Palaeomagnetic field strengths from sediments baked by lava flows of the Chaîne des Puys, France," *Nature*, V. **278**, p. 153-156, 8 Mar 1979.
Bard, E., Hamelin, B., Fairbanks, R. G. and Zinder, A., 1990, "Calibration of the ^{14}C timescale over the past 30,000 years using mass spectrometric U-Th ages from Barbados Corals," *Nature*, Vol. **345**, p. 405-410, 31 May 1990.
Barnola, J. M., Raynaud, D., Korotkevich, Y. S. and Lorius, C., 1987, "Vostok ice core provides 160,000 year record of atmospheric CO_2," *Nature*, Vol. **329**, p. 408-413, 1 Oct 1987.
Baxter, J. and Atkins, T., 1976, *The Fire Came By*.
Beer, Jürg, Oeschger, H., Andrée, M., Bonani, G., Suter, M., Wölfli, W. and Langway, C. C. Jr., 1984, "Temporal Variations in the ^{10}Be Concentration Levels Found in the Dye 3 Ice Core, Greenland," *Annals of Glaciology*, V. **5**, p. 16-17.
Black, D. J., 1967, "Cosmic ray effects and faunal extinctions at geomagnetic field reversals," *Earth and Planet. Sci. Lett.*, Vol. **3**, p. 225-236.
Block, C., and Dams, R., 1975, "Inorganic composition of Belgian coals and coal ashes," *Envir. Sci. Techrol.*, Vol. **9**, p. 146-150.

Bloom, A. L., Broecker, W. S., Chappell, J. M. A., Mathews, R. K. and Mesolella, K. J., 1974, "Quaternary sea-level fluctuations on a tectonic coast: New ^{230}Th/^{234}U dates from the Huon Peninsula, New Guinea," *Quat. Res.,* Vol. **4**, p.185-205.
Bohor, B. F., Foord, E. E., Modreski, P. J. and Triplehorn, D. M., 1984, "Mineralogic Evidence for an Impact Event at the Cretaceous-Tertiary Boundary," *Science,* Vol. **224**, p. 867-869, 25 May 1984.
Böhnel, H., Reismann, N., Jäger, G., Haverkamp, U., Negendank, J. F. W. and Schmincke, H. -U., 1987, "Paleomagnetic investigation of Quaternary West Eifel volcanics (Germany): indication for increased volcanic activity during geomagnetic excursion/event?" *Journal of Geophys.,* Vol. **62**, p. 50-61.
Bonhommet, N.. and Babkine, J. *C.r. hebd séance Acad. Sci. Paris* 264, p. 92-94, 1967.
Brabyn, Howard, 1985, "A Precambrian nuclear reactor," in *The Making of the Earth,* by Richard Fifield, p. 100-104.

Cande, Steven C. and Kent, Dennis V., 1992, "A New Geomagnetic Polarity Time Scale for the Late Cretaceous and Cenozoic," *Journal of Geophys. Res.,* Vol. **97**, No. B10, p. 13,917-13,951, 10 Sep 1992.
Cande, Steven C. and Kent, Dennis V., 1995, "Revised calibration of the geomagnetic polarity timescale for the Late Cretaceous and Cenozoic," *Journal of Geophys. Res.,* Vol. **100**, No. B4, p. 6093-6095, 10 Apr 1995.
Carey, S. Warren, 1976, *The Expanding Earth.*
Carey, S. Warren, 1981, *The Expanding Earth, a symposium.*
Carey, S. Warren, 1989, *Theories of the Earth and Universe.*
Carter, N. L., Officer, C. B. and Drake, C. L., 1990, "Dynamic deformation of quartz and feldspar: clues to causes of some natural crises," *Tectonophysics,* Vol.**171**, p. 373-391.
Champion, D. E., Lanphere, M. A. and Kuntz, M. A., 1988, "Evidence for a new geomagnetic reversal from Lava flows in Idaho: Discussion of short polarity reversals in the Brunhes and late Matuyama polarity chrons," *Journal of Geophys. Res.,* Vol. **93**, p. 11667-11680, 10 Oct 1988.
Chappellaz, J., Barnola, J. M., Raynaud, D., Korotkevich, Y. S. and Lorius, C., 1990, "Ice-core record of atmospheric methane over the past 160,000 years," *Nature,* Vol. **345**, p. 127-131, 10 May 1990.
Chauvin, Annick, Duncan, R. A., Bonhommet, N. and Levi, S., 1989, "Paleointensity of the Earth's magnetic field and K-Ar dating of the Louchardière volcanic flow (central France): new evidence for the Laschamp excursion," *Geophys. Res.Lett.,* Vol. **16**, p. 1189-1192
Christiansen, Robert L., 1984, "Yellowstone Magmatic Evolution: Its Bearing on Understanding Large Volume Explosive Volcanism," in *Explosive Volcanism: Inception, Evolution, and Hazards,* p. 84-95.

Chyba, Christopher F., Thomas, Paul J., Zahnle, Thomas, and Zahnle, Kevin, 1993, "The 1908 Tunguska explosion: atmospheric disruption of a stony asteroid," *Nature*, Vol. **361**, p. 40-44, 7 Jan 1993.
Claeys, Phillipe, Casier, Jean-Georges, and Margolis, Stanley V., 1992, "Microtektites and Mass Extinctions: Evidence for a Late Devonian Asteroid Impact,"*Science,* Vol. **257**, p. 1102-1104, 21 Aug 1992.
Clemens, S. C., Farrell, J. W. and Gromet, L. P., 1993, "Synchronous changes in seawater strontium isotope composition and global climate," *Nature*, Vol. **363**, p. 607-610, 17 Jun 1993
Coe, Robert S., Grommé, S. and Mankinen, E. A., 1984, "Geomagnetic Paleointensities from Excursion Sequences in Lavas on Oahu, Hawaii," *Journal of Geophys. Res.*, Vol. **89**, No. B2, p. 1059-1069, 10 Feb 1984.
Coe, Robert S. and Prévot, Michel, 1989, "Evidence suggesting extremely rapid field variations during a geomagnetic reversal," *Earth and Planet. Sci. Lett.,* Vol. **92**, p. 292-298.
Coe, Robert S., Prévot, Michel, and Camps, P., 1995, "New Evidence for extraordinarily rapid change of the geomagnetic field during a reversal," *Nature*, Vol. **374**, p. 687-692, 20 Apr 1995.
Cohen, Sam, 1983, *The Truth about the Neutron Bomb.*
Colbert, Edwin H., 1968, *The Great Dinosaur Hunters and Their Discoveries.*
Courtillot, Vincent E. and Besse, Jean, 1987, "Magnetic Field Reversals, Polar Wander, and Core-Mantle Coupling," *Science,* Vol. **237**, p. 1140-1147, 4 Sep 1987.
Courtillot, Vincent E., 1990, "A Volcanic Eruption," *Scientific American,* p. 85-92, Oct 1990.
Courtillot, Vincent E., 1999, *Evolutionary Catastrophes; The Science of Mass Extinction.*
Cowan, Clyde, Atluri, C. R. and Libby, W. F., 1965, "Possible anti-matter content of the Tunguska Meteor of 1908," *Nature*, Vol. **206**, p. 861-865, 29 May 1965.
Cowan, George A., 1976, "A Natural Fission Reactor," *Scientific American,* p. 36-47, Jul 1976.
Cox, Allan, 1969, "Geomagnetic reversals," *Science*, Vol. **163**, p. 237-245, 17 Jan 1969. (Also in *Plate Tectonics and Geomagnetic Reversals*, p. 207-220.)
Cox, Allan, 1973, *Plate Tectonics and Geomagnetic Reversals.*
Cox, Allan, Dalrymple, G. Brent, and Doell, R. R., 1967, "Reversals of the Earth's Magnetic Field," *Scientific American*, Vol. **216**, p. 44-54. (Also in *Plate Tectonics and Geomagnetic Reversals*, p. 188-206.)
Cox, Allan, Doell, Richard R. and Dalrymple, G. Brent, 1963, "Geomagnetic polarity epochs and Pleistocene geochronometry," *Nature*, Vol. **198**, p. 1049-1051.

Cox, Allan, Doell, Richard R. and Dalrymple, G. Brent, 1964, "Reversals of the Earth's Magnetic Field," *Science*, Vol. **144**, p. 1537-1543. (Also in *Plate Tecto nics and Geomagnetic Reversals*, p. 169-178.)
Cox, Allan and Hart, Robert Brian, 1986, *Plate Tectonics, How it Works.*
Crain, I. K., 1971, "Possible Direct Causal Relation between Geomagnetic Reversals and Biological Extinctions," *Geol. Soc. Amer. Bull.*, Vol. **82**, p. 2603-2606.
Crain, I. K. and Crain, P. L., 1970, "New stochastic model for geomagnetic reversals," *Nature*, Vol. **228**, p. 39-41.
Crain, I. K., Crain, P. L. and Plant, M. G., 1969, "Long Period Fourier spectrum of geomagnetic reversals," *Nature*, Vol. **223**, p. 283.
Creer, Kenneth M., 1975, "On a tentative correlation between changes in the geomagnetic polarity bias and reversal frequency and the Earth's rotation through Phanerozoic time," in *Growth Rhythms and the History of the Earth's Rotation*, G. D. Rosenberg and S. K. Runcorn, editors, p. 293-318.
Croll, James, 1864, *Climate and Time*. Stratospheric Sources of Nitric Oxide," *Science*, Vol. **189**, p. 457-459, 8 Aug 1975.
Currie, K. L., 1967, "Geological notes on the Carswell circular structure," *Geol. Surv. Can. Pap. 67-32*, p. 1-60, 1967.

Dansgaard, W., Johnsen, S. J., Clausen, H. B., Dahl-Jensen, D., Gundestrup, N. S., Hammer, C. U., Hvidberg, C. S., Steffensen, J. P. Sveinbjörnsdottir, A. E., Jouzel, J. and Bond, G., 1993, "Evidence for general instability of past climate from a 250-kyr ice-core record," *Nature*, Vol. **364**, p. 218-220, 15 Jul 1993.
Dansgaard, W., Johnsen, S. J., Clausen, H. B. and Langway, C. C. Jr., 1971, "Climatic Record Revealed by Camp Century Ice Core," in *Late Cenozoic Glacial Ages*, Karl K. Turekian, editor, p. 37-56.
Dansgaard, W., 1981, "Ice Core studies: Dating the past to find the future," *Nature*, Vol. **290**, p. 360-361, 2 Apr 1981.
Dansgaard, W. and Duplessy, J., 1981, "The Eemian interglacial and its termination," *Boreas*, Vol. **10**, p. 219-228.
Darwin, Charles, 1858, *On the Origin of Species.*
Dawson, Alastair G., 1992, *Ice age earth: late quaternary and climate.*
de Michele, Vincenzo, and Serra, Romano, 1997, "Libyan Desert Silica Glass - A Jewel in the Desert," *Meteorite,* May 1997.
Drake, Charles (See Officer, Charles)

Eldredge, Niles, 1985, *Time Frames.*
Eldredge, Niles and Gould, Stephen Jay, 1972, *Punctuated Equilibria: An Alternative to Phyletic Gradualism.*
Elsasser, Walter, Ney, E. P. and Winckler, J. R., 1956, "Cosmic-ray intensity and geomagnetism," *Nature*, Vol. **178**, p. 1226-1227, 1 Dec 1956.
Erwin, Douglas H., 1994, "The Permo-Triassic extinctions," *Nature*, Vol. **367**, p. 231-236, 20 Jan 1994.

Felix, Robert W., 2005, *Not by Fire but by Ice.*
See also www.iceagenow.com
Firestone, Richard, and Topping, William, "Terrestrial Evidence of a Nuclear Catastrophe in Paleoindian Times," *The Mammoth Trumpet*, 2001.
See also: http://www.sciencenews.org/articles/20070602/fobl.asp
and: http://news.bbc.co.uk/2/hi/science/nature/6676461.stm
and: http://americandaily.com/article/19142
and: http://www.georgehoward.net/thecarolinabayevent.htm
Firestone, Richard, and West, Allen, 2007, presented to the American Geophysical Union's Joint Meeting in Acapulco, Mexico, on May 22, 2007.
Firestone, Richard, West, Allen, and Warwick-Smith, Simon, 2006, *The Cycle of Cosmic Catastrophes.*
Fischer, A. G., 1980, "Gilbert-Bedding rhythms and geochronology," *Geol. Soc.Am. Spec. Pap 183*, p. 93-104.
Fischer, A. G., 1981, "Climatic oscillations in the biosphere," *Biotic Crises in Eco logical and Evolutionary Time,* Nitecki, M. H., editor, p. 103-131.
Froelich, Philip N., 1993, "Ruling in the improbable," *Nature*, Vol. **363**, p. 585-587, 17 Jun 1993.
Fuller, Michael D., in "Ancient Magnetic Reversals: Clues to the Geodynamo." See **Hoffman**, Kenneth A.

Gibson, D. W., 1977, "Upper Cretaceous and Tertiary Coal-Bearing Strata in the Drumheller-Ardley Region, Red Deer River Valley, Alberta," *Geol. Surv. Can.Pap. 76-35.*
Gillot, P. Y., Labeyrie, J., Laj, C., Valladas, G., Guerin, G., Poupeau, G. and Delibrias, G., 1979, "Age of the Laschamp paleomagnetic excursion revisited," *Earth and Planet. Sci. Lett.,* Vol. **42**, p. 444-450, 1979.
Gilmour, I., Russell, S. S., Arden, J. W., Lee, M. R., Francki, A. and Pillinger, C. T., 1992, "Terrestrial Carbon and Nitrogen Isotopic Ratios from Cretaceous-Tertiary Boundary Nanodiamonds," *Science*, V. **258**, p. 1624-1626, 4 Dec 1992.
Gold, Thomas, 1987, *Power from the Earth.*
Gold, Thomas and Dyson, Freeman, 2001, *The Deep Hot Biosphere: The Myth of Fossil Fuels.*
Goldschmidt, Richard, 1940, *The Material basis of evolution.*
Gould, Stephen Jay, 1977, "The Telltale Wishbone," *Natural History*, Vol. **86**, No. 9, p. 26-36, Nov 1977.
Gould, Stephen Jay, 1980, *The Panda's Thumb.*
Gould, Stephen Jay, 1989, *Wonderful Life*
Grieve, R. A. F. and Sharpton, V. L., 1981, "The Cretaceous Tertiary extinction event: a cosmic catastrophe?" *Geos*, Vol. 10, p. 7-9.

Harwood, J. M. and Malin, S. R. C., 1976, "Present trends in the Earth's magnetic field," *Nature*, Vol. **259**, p. 469-471. 12 Feb 1976.

Hays, James D., 1970, "Stratigraphy and Evolutionary Trends of Radiolaria in North Pacific Deep-Sea Sediments," *Geol. Soc. Am.* Memoir 126, p. 185-218.

Hays, James D., 1971, "Faunal Extinctions and Reversals of the Earth's Magnetic Field," *Geol. Soc. Am. Bull.*, Vol. **82**, p. 2433-2447, Sep 1971.

Hays, James D., 1973, "The Ice Age Cometh," *Saturday Review of the Sciences*, Vol. **1**, p. 29-32, Apr 1973.

Hays, James D., Imbrie, John, and Shackleton, N. J., 1976, "Variations in the Earth's Orbit; Pacemaker of the Ice Ages," *Science,* Vol. **194**, p. 1121-1132, 10 Dec 1976.

Hays, James D. and Opdyke, Neil D., 1967, "Antarctic Radiolaria, Magnetic Reversals, and Climatic Change," *Science,* Vol. **158**, p. 1001-1011, 24 Nov 1967.

Hays, James D. and Pitman, Walter C. III, 1973, "Lithospheric Plate Motion, Sea Level Changes and Climatic and Ecological Consequences," *Nature*, Vol. **246**, p. 18-22, 2 Nov 1973.

Hays, James D., Saito, Tsunemasa, Opdyke, Neil D. and Burckle, Lloyd H., 1969, "Pliocene-Pleistocene Sediments of the Equatorial Pacific: Their Paleomagnetic, Biostratigraphic, and Climatic Record," *Geol. Soc. Am. Bull.*, Vol. **80**, p. 1481-1514, Aug 1969.

Hildebrand, Alan R. and Boynton, William V., 1990, "Proximal Cretaceous-Tertiary Boundary Impact Deposits in the Caribbean," *Science*, Vol. **248**, p. 843-847, 18 May 1990.

Hildebrand, Alan R. and Boynton, William V., 1991, "Cretaceous Ground Zero," *Natural History,* Jun 1991.

Hitchcock, C. H., 1865, "The Albert Coal, or Albertite, of New Brunswick," Amer. J. Sc. 2nd Ser: **39**: 267-73.

Hoffman, Kenneth A., 1988, "Ancient Magnetic Reversals: Clues to the Geodynamo," *Scientific American*, May 1988.

Horner, John R., 1988, *Digging Dinosaurs.*

Hsü, Kenneth J., 1986, *The Great Dying.*

Hsü, Kenneth J., He, Q. X., McKenzie, J. A., Weissert, H., Perch-Nielsen, K., Oberhänsli, H., Kelts, K., LaBrecque, J., Tauxe, L., Krähenbühl, U., Percival, S. J. Jr., Wright, R., Karpoff, A. M., Petersen, N., Tucker, P., Poore, R. Z., Gombos, A. M., Pisciotto, K., Carman, M. F. Jr., Schreiber, E., 1982, "Mass Mortality and Its Environmental and Evolutionary Consequences," *Science*, Vol. **216**, p. 249-256, 16 Apr 1982.

Hsü, Kenneth J., McKenzie, J. A. and He, Q. X., 1982, "Terminal Cretaceous environmental and evolutionary changes, *Geol. Soc. Am. Spec. Pap. **190***, p. 317-328.

Hsü, Kenneth J., Montadert, L., Bernoulli, D., Cita, Maria Bianca, Erickson, A., Garrison, R. E., Kidd, R. B., Mèlierés, F. Müller, C. and Wright, Ramil, 1977, "History of the Mediterranean salinity crisis," *Nature*, Vol. **267**, p. 399-403, 2 Jun 1977.

Jackson, A. A. and Ryan, M. P., 1973, "Was the Tungus Event due to a Black Hole?" *Nature*, Vol. **245**, p. 88-89, 14 Sept 1973.

Johnson, Kimberly, 2008, "Earth's Core, Magnetic Field Changing Fast, Study Says," *National Geographic News*, 30 Jun 2008. http://news.nationalgeographic.com/news/pf/76158139.html

Keller, Gerta, and Adatte, Thiery, 2007, presenting at the annual meeting of the Geological Society of America in Denver, Coloradeo, 31 Oct 2007.

Kennett, James P., 1977, "Cenozoic evolution of Antarctic glaciation, the CircumAntarctic Ocean, and their impact on global paleooceanography," *Journal of Geophys. Res.*, Vol. **82**, p. 3843-3860, 20 Sep 1977.

Kennett, James P., McBirney, A. R. and Thunell, Robert C., 1977, "Episodes of Cenozoic volcanism in the circum-Pacific region," *Journal of Volcanology and Geothermal Research*, Vol. **2**, p. 145-163.

Kennett, James P. and Thunell, Robert C., 1975, "Global Increase in Quaternary Explosive Volcanism," *Science*, Vol. **187**, p. 497-503, 14 Feb 1975.

Kennett, James P. and Thunell, Robert C., 1977, "Comments on Cenozoic Explosive Volcanism Related to East and Southeast Asian Arcs," in "Island Arcs, deep sea trenches, and back-arc basins," Talwani and Pitman, editors, *American Geophysical Union Maurice Ewing Series*, No. **1**, p. 348-352.

Kennett, James P. and Watkins, N. D., 1970, "Geomagnetic polarity change, volcanic maxima and faunal extinction in the south Pacific," *Nature*, Vol. **227**, p. 930-934, 29 Aug 1970.

Kent, Dennis V. and Gradstein, Felix M., 1986, "A Jurassic to recent chronology," in *The Geology of North America*, Vol. **M**, *The Western North Atlantic Region,* edited by P. R. Vogt and B. E. Tucholke, p. 45-50.

Kent, Dennis V. and Opdyke, N. D., 1976, "Relative paleomagnetic field intensity variations from deep-sea and sediment cores," *Abstract Transactions Am. Geophys. Union*, Vol. **57**, p. 237, 1976.

Kent, Dennis V. and Opdyke, N. D., 1977, "Paleomagnetic field intensity variation recorded in a Brunhes Epoch deep-sea sediment core," *Nature*, Vol. **266**, p. 156-159, 10 Mar 1977.

Kopper, John S., 1976, "Dating and Interpretation of Archeological Cave Deposits by the Paleomagnetic Method," Doctoral Thesis, Columbia University.

Kucha, H., 1981, "Precious metal alloys and organic matter in the Zechstein copper deposits, Poland," *TMPM Tschermaks Min. Petr. Mitt.*, Vol. **28**, p. 1-16.

Kukla, George J., 1975, "Missing link between Milankovitch and climate," *Nature*, Vol. **253**, p. 600-603, 20 Feb 1975.
Kukla, George J., Berger, A., Lotti, R. and Brown, J., 1981, "Orbital signature of interglacials," *Nature*, Vol. **290**, p. 295-300, 26 Mar 1981.
Kukla, George J. and Matthews, Robley K., 1972, "When will the present interglacial end?" *Science*, Vol. **178**, p. 190-191, 13 Oct 1972.
Kukla, George J., Matthews, R. K. and Mitchel, J. M. Jr., 1972, "The end of the Present Interglacial," *Quat. Res.*, Vol. **2**, p. 261-269.
Kurtén, Björn, 1981, *How to Deep-Freeze a Mammoth*.

Labandeira, C. C. and Sepkoski, J. J. Jr., 1993, "Insect diversity in the fossil record," *Science*, 16 Jul 1993.
Laj, Carlo, Guitton, Sylvie, Kissel, Catherine, and Mazaud, Alain, 1988, "Complex behavior of the geomagnetic field during three successive polarity reversals, 11-12 m.y.B.P.," *Journal of Geophys. Res.*, Vol. **93**, p. 11,655-11,666, 10 Oct 1988.
Laj, Carlo, Mazaud, Alain, Weeks, Robin, Fuller, Mike, and Herrero-Bervera, Emilio, 1991, "Geomagnetic reversal paths," *Nature*, Vol. **351**, p. 447, 6 Jun 1991.
Lamb, H. H., 1972, *Climate, Present, Past, and Future*, Vol. 1, p. 432.
Lamb, H. H., 1982, *Climate, history and the modern world*.
Larson, Roger L. and Pitman, Walter C. III, 1972, "World-wide correlation of Mesozoic Magnetic Anomalies, and its Implications," *Geol. Soc. Am. Bull.*, Vol. **83**, p. 3645-3662. (Also in *Megacycles: Long-Term Episodicity in Earth and Planetary History*.)
Libby, Willard F., 1981, *Talking to People*.
Liddicoat, J. C., 1992, "Mono Lake Excursion in Mono Basin, California, and at Carson Sink and Pyramid Lake, Nevada," *Geophys. J. Int.*, V. **108**, p. 442-452.
Liddicoat, J. C. and Coe, R. S., 1979, "Mono Lake geomagnetic excursion," *Journal of Geophys. Res.*, Vol. **84**, p. 261-271, 10 Jan 1979.
Lin, Robert, 2004, as quoted by Suplee, Curt, 2004, "The Sun: Living With a Stormy Star," *National Geographic*, July 2004, p.1-33.
Loper, David E. and McCartney, Kevin, 1986, "Mantle Plumes and the Periodicity of Magnetic Field Reversals," *Geophys. Res. Lett.*, Vol. **13**, No. 13, p. 1525-1528, Dec 1986.
Loper, David E., McCartney, Kevin, and Buzyna, George, 1988, "A Model of Correlated Episodicity in Magnetic-Field Reversals, Climate, and Mass Extinctions," *Journal of Geology*, Vol. **96**, p. 1-15.
Lorenz, R. D. *et al.*, "Titan's inventory of organic surface materials, "*Geophysical Research Letter*s, **35**, 29 Jan 2008.
Lowrie, W., 1989, "Magnetic time scales and reversal frequency," in *Geomagnetism and Palaeomagnetics,*" ed. F. J. Lowes *et al*, p. 155-183.

Lyell, Charles, 1843, "On the upright Fossil Trees found at different levels in the Coal Strata of Cumberland, Nova Scotia," *American Journal of Science*, 1:45, p. 353-356.

Lyons, J. B. and Officer, Charles B., 1992, "Mineralogy and petrology of the Haiti Cretaceous/Tertiary section," *Earth Planet. Sci. Lett.*, Vol. **109**, p. 205-224.

Malin, S. R. C. and Clark, Anne D., 1974, "Geomagnetic Secular Variation, 1962.5 to 1967.5," *Journ. Royal Astro. Soc.* Vol. **36**, p. 11-20.

Malkus, W. V. R., 1968, "Precession of the Earth as the Cause of Geomagnetism," *Science*, Vol. **160**, p. 259-264, 19 Apr 1968.

Mandea, Mioara, see Olsen, Nils

Mankinen, Edward A. and Wentworth, Carl M., "Preliminary Paleomagnetic Results from the Coyote Creek Outdoor Classroom Drill Hole," Santa Clara Valley, California, USGS Open File Report 03-187 2003.

Matuyama, Motonori, 1929, "On the Direction of Magnetisation of Basalt in Japan, Tyôsen and Manchuria," *Japan Academy Proceedings*, Vol. **5**, p. 203-205. (Also in *Plate Tectonics and Geomagnetic Reversals*, p. 154-156.)

Mazaud, A., Laj, C., Bard, E. and Arnold, M., 1992, "Geomagnetic Calibration of the Radiocarbon time-scale," in *The Last Deglaciation: Absolute and Radiocarbon Chronologies*, (eds. Bard, E. & Broecker, W. S.) NATO ASI Series 2, 1992.

Mazaud, A., Laj, C., Laurent de Sèze and Verosub, K. L., 1983, "15-Myr periodicity in the frequency of geomagnetic reversals since 100myr," *Nature*, Vol. **304**, p. 328-330, 28 Jul 1983.

Mazaud, A., Laj, C., Laurent de Sèze and Verosub, K. L., 1984, Letter to *Nature*, Vol. **311**, 27 Sept 1984.

McElhinny, M. W., 1971, "Geomagnetic Reversals during the Phanerozoic," *Science*, Vol. **172**, p. 157-159.

McElhinny, M. W. and Senanayaka, W. E., 1982, "Variations in the Geomagnetic Dipole 1: The Past 50,000 years," *J. Geomagn. Geoelec. Kyoto*, V. **34**, p. 39-51.

McFadden, Phillip L. and Merrill, Ronald T., 1995, "History of Earth's magnetic field and possible connections to core-mantle boundary processes," *Journal of Geophys. Res.*, Vol. **100**, No. B1, p. 307-316, 10 Jan 1995.

McFadden, P. L., Merrill, R. T. and McElhinny, M. W., 1988, "Dipole/Quadrupole Family Modeling of Paleosecular Variation," *Journal of Geophys. Res.*, Vol. **93**, p. 11,583-11,588, 10 Oct 1988.

McLaren, Digby J., 1982, "Frasnian-Famennian extinctions," *Geol. Soc. Am. Spec. Pap. 190*, p. 477-484.

Megacycles*: Long-Term Episodicity in Earth and Planetary History*, 1981, editor, George E. Williams.

Melosh, H. J., 1992, "Airblast scars on Venus," *Nature*, Vol. **358**, p. 622-623, 20 Aug 1992.

Melosh, H. J., 1993, "Tunguska comes down to Earth," *Nature*, Vol. **361**, p. 14-15, 7 Jan 1993.
Merrill, Ronald T. and McElhinny, M. W., 1983, *The Earth's Magnetic Field.*
Merrill, Ronald T. and McFadden, P. L., 1990, "Paleomagnetism and the Nature of the Geodynamo," *Science*, Vol. **248**, p. 345-350, 20 Apr 1990.
Merrill, Ronald T. and McFadden, 1995, "Dynamo theory and paleomagnetism," *Journal of Geophys. Res.*, Vol. **100**, No. B1, P. 317-326, 10 Jan 1995.
Mörner, Nils-Axel, 1971, "The Plum Point Interstadial: Age, Climate, and Subdivision," *Canadian Journal Earth. Sciences,* Vol. **8**, p. 1423-1431.
Mörner, Nils-Axel and Lanser, Johan P., 1974, "Gothenburg magnetic 'flip'," *Nature*, Vol. **251**, p. 705-706, 4 Oct 1974.
Mörner, Nils-Axel, Lanser, Johan P. and Hospers, J., 1971, "Late Weichselian Palaeomagnetic Reversal," *Nature Physical Science*, Vol. **234**, p. 173-174, 27 Dec 1971.

Neftel, A., Oeschger, H., Stafflebach, T. and Stauffer, B., 1988, "CO_2 record in the Byrd ice core 50,000-5,000 year BP," *Nature*, Vol. **231**, p. 609-611, 18 Feb 1988.
Negi, J. G. and Tiwari, R. K., 1983, "Matching Long Term Periodicities of Geomagnetic Reversals and Galactic Motions of the Solar System," *Geophys. Res. Lett.,* Vol. **10**, No. 8, p. 713-716, Aug 1983.
Negi, J. G. and Tiwari, R. K., 1984, "Periodicities of palaeomagnetic intensity and palaeoclimatic variations: a Walsh spectral approach," *Earth and Planet. Sci. Lett.,* Vol. **70**, p. 139-147, 1984.
Nellis, W. J., Ree, F. H., van Thiel, M. and Mitchell, A. C., 1981, "Shock compression of liquid carbon monoxide and methane to 90 Gpa (900kbar)[a], *Journal of Chemical Physics,* Vol. **75**, p. 3055-3063.
Noyes, Robert W., 1982, *The Sun, Our Star.*

Officer, Charles B., 1990, "Extinctions, iridium and shocked minerals associated with the Cretaceous/Tertiary Transition," *Journal Geological Education*, Vol. **38**, p. 402-425.
Officer, Charles B. and Carter, N. L., 1991, "A review of the structure, petrology and dynamic deformation characteristics of some enigmatic terrestrial structures," *Earth-Science Review*, Vol. **30**, p. 1-49.
Officer, Charles B. and Drake, Charles L., 1985, "Epeirogeny on a Short Geological Time Scale," *Tectonics,* Vol. **4**, p. 603-612.
Officer, Charles B. and Drake, Charles L., 1983, "The Cretaceous-Tertiary Transition," *Science*, Vol. **219**, p. 1383-1390, 25 Mar 1983.
Officer, Charles B. and Drake, Charles L., 1985, "Terminal Cretaceous Environmental Events," *Science,* 8 Mar 1985, Vol. **227**, p. 1161-1166.
Officer, Charles B. and Drake, Charles L., 1985, answer to Smit, Kyte, and French (in Letters), *Science*, Vol. **230**, 13 Dec 1985.

Officer, Charles B., Hallam, Anthony, Drake, Charles L. and Devine, Joseph D., 1987, "Late Cretaceous and paroxysmal Cretaceous/Tertiary extinctions," *Nature* Vol. **326**, p. 143-149, 12 Mar 1987.
Olsen, Nils and Mandea, Mioara, "Rapidly changing flows in the Earth's core," *Nature Geoscience* **1**, 18 May 2008, p. 390-394.

Padian, Kevin, and Faux, Cynthia Marshal, The opisthotonic posture of vertebrate skeletons: postmortem contraction or death throes?, *Journal of Paleobiology*, Mar 2007. Also in the *San Francisco Chronicle*, 6 Jun 07.
Pagel, M., Wheatley, K. and Ey, F., 1985, "The Origin of Carswell Structure," *Geol. Soc. Can. Spec. Pap. 29*, p. 213-223.
Pal, P. C. and Creer, K. M., 1986, "Geomagnetic reversal spurts and episodes of extraterrestrial catastrophism," *Nature,* Vol. **320**, p. 148-150, 13 Mar 1986.
Pederson, K. R. and Lam, J., 1970, "Precambrian organic compounds from the Ketilidian of south-western Greenland," *Greenlands Geologiske Unders. Bull.* No. 82.
Peterman, Zell E., Hedge, Carl E. and Tourtelot, Harry A., 1970, "Isotopic composition of strontium in sea water throughout Phanerozoic time," *Geochim. et Cosmochim. Acta* Vol. **34**, p. 105-108 and 111-118. (Also in *Megacycles: Long-Term Episodicity in Earth and Planetary History*.)
Prévot, Michel, Mankinen, E. A., Coe, R. S. and Grommé, C. S., 1985, "The Steens Mountain (Oregon) Geomagnetic Polarity Transition 2. Field Intensity Variations and Discussions of Reversal Models," *Journal of Geophys. Res.,* Vol. **90**, p. 10,417-10,448, 10 Oct 1985.

Raisbeck, G. M., Yiou, F., Bourles, D., Lorius, C., Jouzel, J. and Barkov, N. I., 1987, "Evidence for two intervals of enhanced ^{10}Be deposition in Antarctic ice during the last glacial period," *Nature*, Vol. **326**, p. 273-277, 19 Mar 1987.
Raisbeck, G. M., Yiou, F., Fruneau, M., Loiseaux, J. M., Lieuvin, M., Ravel, J. C. and Hays, J. D., 1979, "A search in a marine sediment core for ^{10}Be concentration variations during a geomagnetic field reversal," *Geophys. Res. Lett.*, p. 717-719, Sep 1979.
Rampino, Michael R., 1979, "Possible relationships between changes in global ice volume, geomagnetic excursions, and the eccentricity of the Earth's orbit," *Geology*, Vol. **7**, p. 584-587, Dec 1979.
Rampino, Michael R., 1981, "Revised age estimates of Brunhes palaeomagnetic events: Support of a link between geomagnetism and orbital eccentricity variations," *Geophys. Res. Lett.*, Vol. **8**, p. 1047-1050.
Rampino, Michael R. and Reynolds, Robert C., 1983, "Clay Mineralogy of the Cretaceous-Tertiary Boundary Clay," *Science*, Vol. **219**, p. 495-498, 4 Feb 1983.
Rampino, Michael R. and Stothers, Richard B., 1984, "Terrestrial mass extinctions, cometary impacts and the Sun's motion perpendicular to the galactic plane," *Nature,* Vol. **308**, p. 709-712, 19 Apr 1984.

Rampino, Michael R. and Stothers, Richard B., 1984, "Geological Rhythms and Cometary Impacts," *Science*, Vol. **226**, p. 1427-1431, 21 Dec 1984.
Rampino, Michael R. and Stothers, Richard B., 1988, "Flood Basalt Volcanism During the past 250 Million Years," *Science*, Vol. **241**, p. 663-668, 5 Aug 1988.
Raup, David M. and Sepkoski, J. John Jr., 1984, "Periodicity of extinctions in the geologic past," *Proc. Natl. Acad. Sci.*, Vol. **81**, p. 801-805, Feb 1984.
Raup, David M. and Stanley, Steven M., 1978, *Principles of Paleontology*.
Raup, David M., 1985, "Magnetic Reversals and mass extinctions," *Nature*, Vol. **314**, p. 341-343, 28 Mar 1985.
Raup, David M., 1986, *The Nemesis Affair*.
Raup, David M., 1991, *Extinction: Bad Genes or Bad Luck?*
Raup, David M. and Jablonski, D., 1993, "Geography of End-Cretaceous Marine Bivalve Extinctions," *Science*, Vol. **260**, p. 971-973, 14 May 1993.
Ray, Dixy Lee with Guzzo, Louis R., 1990, *Trashing the Planet*.
Ray, Dixy Lee with Guzzo, Louis R., 1993, *Environmental Overkill, Whatever Happened to Common Sense?*
Roperch, P., Bonhommet, N. and Levi, S., 1988, "Paleointensity of the earth's magnetic field during the Laschamp excursion and its geomagnetic implications," *Earth and Planet. Sci. Lett.*, Vol. **88**, p. 209-219, 1988.
Ryan, W. B. F. and Cita, M. B., 1977, "Ignorance concerning episodes of ocean-wide stagnation," *Mar. Geol.*, Vol. **23**, p. 197-215.

Sagan, Carl, Toon, O. and Gierash, P., 1973, "Climatic change on Mars," *Science*, Vol. **181**, p. 1045-1049.
Sagan, Carl, 1991, "Titan: Key to the Origins of Life?" *Parade Magazine*, p. 6-11, 1 Dec 1991.
Sagan, Carl, 1991, "When Worlds Collide," *Parade Magazine*, p. 4-6, 3 Mar 1991.
Sepkoski, J. John Jr., 1982, "Mass Extinctions in the Phanerozoic oceans: A review," *Geol. Soc. Am. Spec. Pap. 190*, p. 283-289.
Scott, Donald E., 2006, *The Electric Sky*.
Sharpton, Virgil L., Burke, K., Camargo-Zanoguera, A., Hall, S. A., Lee, D. S., Marín, L. E., Suárez-Reynoso, G., Quezada-Muñeton, J. M., Spudis, P. D. and Urratia-Fucugauchi, Jaime, 1993, "Chicxulub Multiring Impact Basin: Size and other Characteristics Derived from Gravity Analysis," *Science*, Vol. **261**, p 1564-1567, 17 Sep 1993.
Simpson, John F., 1966, "Short Notes, Evolutionary Pulsations and Geomagnetic Polarity," *Geol. Soc. Am. Bull.*, Vol. **77**, p. 197-204, Feb 1966.
Smit, Jan, 1982, "Extinction and evolution of planktonic foraminifera after a major impact at the Cretaceous/Tertiary boundary," *Geol. Soc. Am. Spec. Pap. 190*, p. 329-352.
Smit, Jan and van der Kaars, S., 1984, "Terminal Cretaceous extinctions in the Hell Creek area: compatible with catastrophic extinctions," *Science*, Vol. **223**, p. 1177-1179, 16 Mar 1984.

Smith, Peter J., editor, 1986, *The Earth*.
Stanley, Steven M., 1987, *Extinction*.
Stanley, Steven M., 1984, "Mass Extinctions in the Ocean," *Scientific American*, p. 64-72, June 1984.
Stanley, Steven M., 1984, "Temperature and biotic crises in the marine realm," *Geology*, Vol. **12**, p. 205-208, Apr 1984.
Stanley, Steven M., 1988, "Paleozoic Mass Extinctions: Shared Patterns Suggest Global Cooling as a Common Cause," *Am. Journal of Sci.*, Vol. **288**, p. 334-352.
Steel, Duncan, "Tunguska at 100," *Nature*, Vol. **453**, 26 Jun 2008, p. 1157-1159
Steiner, J., 1973, "Possible galactic causes for synchronous sedimentation sequences of the North American and eastern European cratons," *Geology*, Vol. **1**, p. 89-92.
Steiner, J., 1977, "An expanding Earth on the basis of sea-floor spreading and subduction rates," *Geology*, Vol. **5**, p. 313-318, May 1977.
Steiner, J., 1978, "Lead isotope events of the Canadian shield, *ad hoc* solar galactic orbits and glaciations," *Precambrian Research*, Vol. **6**, p. 269-274.
Steiner, J., and Grillmair, E., 1973, "Possible Galactic Causes for Periodic and Episodic Glaciations," *Geol. Soc. Am. Bull.,* Vol. **84**, Mar 1973, p. 1003-1018.
Stoner, Joseph, Spoken at an American Geophysical Union meeting, Dec 2005.
Stothers, Richard B., 1986, "Periodicity of the Earth's magnetic reversals," *Nature,* Vol. **322**, p. 444-446, 31 Jul 1986.
Suplee, Curt, 2004, "The Sun: Living With a Stormy Star," *National Geographic*, July 2004, p.1-33.

Thomas, Rob, 1993, Personal Communication
Tschudy, R. H., Pellmore, C. L., Orth, C. J., Gilmore, J. S. and Knight, J. D., 1984, "Disruption of the Terrestrial Plant Ecosystem at the Cretaceous-Tertiary Boundary, Western Interior," *Science*, Vol. **225**, p. 1030-1032.
Uffen, Robert J., 1963, "Influence of the Earth's Core on the Origin and Evolution of Life," *Nature,* Vol. **198**, p. 143-144, 13 Apr 1963.

Valet, J. P., Laj, C. and Langereis, C. G., 1988, "Sequential geomagnetic reversals recorded in upper Tortonian marine clays in western Crete (Greece)," *Journal of Geophys. Res.* Vol. **93**, p. 1131-1151, 1988.
Velikovsky, Immanuel, 1950, *Worlds in Collision*.
Velikovsky, Immanuel, 1955, *Earth in Upheaval*
Verosub, K. L. and Banarjes, S. K., 1977, "Geomagnetic excursions and their paleomagnetic record," *Rev. Geophysics and Space Physics,* Vol. **15**, p. 145-155.

Vine, Fred J., 1966, "Spreading of the Ocean Floor: New Evidence," *Science*, Vol. **154**, p. 1405-1415, 16 Dec 1966. (Also in *Plate Tectonics and Geomagnetic Reversals*, p. 245-264.)
Vogt, Peter R., 1975, "Changes in Geomagnetic Reversal Frequency at times of Tectonic Change: Evidence for Coupling between core and upper mantle processes," *Earth and Planet. Sci. Lett.*, Vol. **25**, p. 313-321.

Waddington, C. J., 1967, "Paleomagnetic field reversals and cosmic radiation," *Science*, Vol. **158**, p. 913-915.
Ward, Peter D., 1992, *On Methuselah's Trail*.
Watkins, N. D. and Goodell, H. G., 1967, "Geomagnetic Polarity Change and Faunal Extinction in the Southern Ocean," *Science,* Vol. **156**, p. 1083-1087, 26 May 1967.
Wezel, F. C., Vannucci, S. and Vannucci, R., 1981, "Decouverte de divers niveaux riches en iridium dans la "Scaglia Rossa" et la "Scaglis bianca" de l'Appennin d'Ombrie-Marches (Italie)," *C.R. Acad. Sci. Paris,* Sér. II, 293, p. 837-844.
Winkler, Gisel, 2005, www.sciencedaily.com/released/2005/04/050428175700
Wolbach, Wendy S., Lewis, Roy S. and Anders, Edward, 1985, "Cretaceous Extinctions: Evidence for Wildfires and Search for Meteoritic Material," *Science,* Vol. **230**, p. 167-170, 11 Oct 1985.
Wright, Giles, 1999, "The Riddle of the Sands," *New Scientist*, p. 42-45, 10 July 1999.

Yiou, F.; Raisbeck, G. M., Baumgartner, S., Johnsen, S., Jouzel, J., Kubik, P.W., Lestringuez, J., Stiésvenard, M., Suter, M,, Yiou, P., "Beryllium 10 in the Greenland Ice Core Project ice core at Summit, Greenland," *Jour. of Geophys. Res.* Vol. **102**, Issue C12, p. 26783-26794. 00/1997.

Zimmer, Carl, 2001, *Evolution: The Triumph of an Idea*.
Zolotov, A.V., 1969, "The Problem of the Tunguska Catastrophe of 1908, Minsk: *Naukai.Tekhnika*, p. 74, 110, 118, 1969.

INDEX

Age of the Fishes, 25
Afar specimen, 70
Ager, Derek V., 27
Allaby, Michael, 47, 130
Allosaur, 30
Alvarez, Walter, 45-47, 129
Ammonites, 25, 28, 30, 31, 56, 62, 72
Angelopoulos, Vassilis, 155
Angiosperm, 30-32
Anophthalmia, 61
Anoxic waters, 43, 138
Anthracite, 127
Apennine Mountains, 45
Apes, 73
Aptian, 43
Asphalt, 124, 125
Asteroid, 113, 152, 163
Athena, 71
Atomic, 63, 81-84,
Atomic bomb, 105, 116, 117
Atomic Energy Commission, 49, 54, 115
Atomic particles, 82, 83
Aurora borealis (see Northern Lights)
Australopithecus, 73
Axis of rotation, 89, 90, 93

Baadsgaard, H., 157, 160
Bakker, Robert, 20, 29, 30, 56, 57, 58, 61, 65, 68, 71
Babkine, J., 98
Barbados, 102
Barbetti, M., 99
Bard, Edouard, 52

Barosaurus, 66, 68
Barnola, J. M., 52
Basalt, 50
Baxter, J., 107, 117
Beer, Jürg, 53
Berggren, William, 33, 34
Bernissart, 155
Beryllium-10, 49, 53. 54, 102-104, 115, 162, 163
Big Bang, 132
Big Eloise, 147
Birds, 33, 58, 59, 66, 68
Bitumen, 135, 165
Bituminous coal, 117
Biwa I, Biwa II, Biwa III, 100, 101
Black mat, 146-148
Black muds, shales, 43, 44, 124, 137-140, 160, 162
Blake magnetic reversal, 73, 100-102
Block, C., 111
Bloom, A. L., 103
Bohor, Bruce, 160
Bonhommet, Norbert, 98
Bosumtri Crater, 108
Bow shock, 154
Brabyn, Howard, 140
Brachiopod, 33
Brontosaurus, 29, 67
Brown coal, 127, 128, 131
Brunhes, 50, 53, 73
Brunhes/Matuyama boundary, 50, 97, 100
Bryozoan, 111
Buckyballs, 150

Calcium, 63, 64, 85
Callovian, 110
Cambrian, 24, 25, 43
Camps, P., 96
Cande, Steven, 99
Carbon, 24, 43, 44, 52-54, 111, 112, 114-116, 122-126, 133, 135-140, 144-153, 156-160
Carbon cycle, 52
Carbon dioxide, 52, 104
Carbon mat, 146-148
Carbon^{-14}, 49, 52, 54, 84, 102-104, 114, 116, 117, 162, 163
Carboniferous, 25, 26, 138
Carey, S. Warren, 39, 105, 119, 140
Carolina Bays, 104, 150-152, 156, 166
Carswell Structure, 141, 142
Carter, Neville, 142
Casier, Jean-Georges, 110
Cassini, 134
Caucacus Mountains, 107
Celestial orbit, 89, 95
Cenomanian, 43
Cenozoic, 101
Ceratid, 28
Chaîne des Puys, 98
Champion, D. E., 101
Chappellaz, J., 53
Chert, 148, 149
Chicxulub Crater, 108, 109, 120
Christiansen, Robert, 103
Chyba, Christopher, 108
Claeys, Phillipe, 109
Clemens, S. C., 52
Clovis, 147-150

Coal, 26, 44, 111, 124-128, 131, 133, 137, 143, 158-160, 162
Cockroaches, 26
Coe, R. S., 95
Cohen, Sam, 116
Colbert, Edwin H., 158
Como Bluff, Wyoming, 29
Cosmic rays, 46, 49, 62, 82, 114, 115
Cosmic year, 88
Courtillot, Vincent, 72, 99, 111, 112
Cowan, Clyde, 116, 117
Cowan, George, 140
Cox, Allan, 45, 92
Creation, 30, 64, 87

Danish National Space Center, 94
deHeer, Walt, 137
deMichele, Vincenzo, 121
Devonian, 25, 43, 109, 110, 138
Diamonds, 143-146, 148-150, 157-159, 162, 163
Dinosaur, 29-32, 40, 41, 44, 56-59, 65-68, 75, 108, 114, 120, 122, 128, 131, 143, 144, 152, 158-162, 164
DNA, 60
Drake, Charles, 42
Dryan, Ann, 87
Duckbill dinosaur, 57, 58

Einstein, Albert, 85, 122
Eldredge, Niles, 21, 73
Electromagnetic, 62, 130, 132
Elsasser, Walter, 92
Emu, 69

Eocene, 57, 109, 110
Equinoctial precession (see precession of the equinoxes)
Erwin, Douglas, 118
Evolution, 19-21, 29, 35-38, 49, 57, 59, 60, 72, 87, 160
Exencephaly, 61
Extinction, 20, 24, 25, 28, 29, 31, 35-40, 44, 46, 47, 49, 54-58, 72, 101, 103, 104, 108-113, 116-118, 129, 130, 138, 146, 150, 157, 160, 164

Falconer, Hugh, 20
Fallout shelter, 166
Faux, Cynthia, 41
Ferguson Ranch, 110
Fern spike, 31
Firestone, Richard, 103, 104, 145, 149-154, 166
Fischer, Alfred, 43
Fish Clay, 34, 43, 45, 120
Frasnian-Famennian, 43
Foraminifera, 31, 34
Froelich, Philip, 51, 52
Fuller, Buckminster, 150
Fuller, Michael D., 96
Fullerenes, 150

Gabon, 140
Galactic orbit, 88
Gamma ray, 61
Gas, 156, 159
Gastrolith, 68
Gauss, 79
Geologic record, 51
Geologic time scale, 22, 23
Geomagnetic reversal (see magnetic reversal)
General Electric, 157

Geraldine Bone Bed, 156
German Research Center for Geoscience, 94
Gibson, D. W., 126
Gillette, David, 68
Gilmour, Iain, 124, 144
Giraffe, 69, 70
Gizzard, 68
Glaciation, 88, 101
Goddard Space Flight Center, 154
Gold, Thomas, 126, 131, 133, 136, 143, 160, 161
Goldschmidt, Richard, 59, 60
Google Earth, 151
Gothenburg, 53, 98, 100, 101, 103, 104, 152, 165
Gould, Stephen Jay, 20, 21, 31, 36, 37, 55, 70, 71, 90
Grady ponds, 150
Graphite, 157, 159
Graptolite, 28
Greenland Space Science Symposium, 92
Grieve, R. A. F., 116
Grillmair, E., 88
Gubbio, 34, 45, 46, 49
Grommé, C. S., 95
Gulf of Mexico, 104
Guzzo, Lou, 54

Hannay, James, 157
Hart, Robert, 45
Harwood, J. M., 93
Haynes, Vance, 147
Hays, James D., 72
Helium, 76, 112, 131, 140
Helium-3, 49, 115
Hell Creek Formation, 159
Herron, M. M., 53

Hildebrand, Alan, 109
Hiroshima, 61, 105, 106, 117, 120, 129
Hitchcock, G. H., 127
Hoffman, Kenneth A., 96
Holocene, 98
Homo erectus, 73
Homo habilis, 73
Hopeful monster, 59, 72
Horner, Jack, 65
Horse, 35, 36, 65, 67, 68, 70, 103, 147
Hsü, Kenneth J., 28, 42
Hunstville, Alabama, 166
Huxley, Thomas, 20
Hydrocarbon, 54, 134-136, 159, 162
Hydrocephaly, 61
Hydrogenation, 159

Ice age, 37, 51, 52-54, 88, 89, 98, 101-105, 146
Ichtyostega, 25
Igneous rocks, 50
Iguanodon, 156
Insects, 25, 26, 28
Ionization, 82, 84, 130
Iridium, 45, 46, 73, 104, 108, 110-113, 116-118, 120-122, 146, 153, 160-162
Iron balls (iron missiles), 148, 149, 153, 156

Jaramillo, 100
Java man, 73
Johnson, Kimberly, 194
Joules, 155
Jupiter, 135, 136, 160
Jurassic, 29, 30, 109

Keller, Gerta, 35
Kennett, James P., 48, 102, 146
Kent, Dennis V., 125
Kerogen, 111, 124, 139
Kinnekulle, 138
Kohm shales, 140
Kopper, John S., 53, 54, 98
K-T boundary, 33, 40, 41, 45, 46, 56, 57, 109-113, 117, 118, 120, 123, 124, 137, 143-145, 157, 159, 160, 164
Kucha, H., 111
Kukla, George, 99
Kurtén, Björn, 125

Labandeira, C. C., 28
Laetoli specimen, 71
Lake Bonneville, 103, 104
Lake Missoula, 102-104
Lake Mungo, 53, 73, 99, 100, 101, 103, 104, 153, 165
Lamb, H. H., 104
Lamont-Doherty Earth Observatory, 52, 72, 99, 116, 138
Lampshells, 33
Lanser, Johan P., 98, 152
Laschamp, 98-100, 103, 154
Lawrence Berkeley National Laboratory, 148
Lawrence Livermore National Laboratory, 156
Lead, 88
Lerbekmo, J. F., 160
Libby, Willard F., 49, 106, 107, 115, 115, 116, 135
Libyan Desert, 121, 150, 156
Liddicoat, J. C., 99

INDEX 187

Lignite, 124, 127, 128, 131, 160
Lorenz, Ralph, 135
Lovelock, James, 47
Lumberton, N.C., 151
Lyell, Charles, 126
Lyons, James, 142

Macromutation, 59, 60
Magnetic
 bow shock, 154
 excursion, 21, 50, 51, 53, 95-101, 103, 104, 136, 152, 153, 163, 165
 field, 46-50, 62, 72, 74, 76-79, 82, 83, 85, 91-96, 98, 101, 104, 106, 132, 136, 153, 155, 163, 164, 166
 fluctuations, 92-96, 99
 intensity, 94, 95, 98, 99, 136
 reversal, 21, 45, 46-51, 53, 54, 60, 72-75, 79, 82-85, 87, 88, 92-95, 97-104, 136, 156, 162, 167
 ropes, 154, 155
 shield, 156
Magnetite, 49, 50
Magnetopause, 155
Magnetosphere, 47, 80, 81
Magnetostratigraphy, 49
Malin, S. R. C., 93
Mammoth, 37, 54, 69, 103, 104, 125, 146-150, 153, 157, 165
Mandea, Mioara, 194
Manhattan project, 115
Mankinen, Edward A., 51, 95, 98

Manson Structure, 142
Mars, 90
Maryland basins, 150
Mastodon, 37, 146, 147, 165
Matuyama, 50, 97, 100
Mazaud, Alain, 52
McFadden, P. L., 93
McElhinny, M. W., 93
McLaren, Digby, 43
Merrill, Ronald T., 93
Mesozoic, 138
Meteor, 46, 107, 108, 110, 112, 113, 117, 120, 121, 123, 124, 134, 135, 141, 144, 150, 152, 163
Meteorite, 152
Methane, 53, 54, 133-135, 159
Michigan, 149
Microcephaly, 61
Miocene, 73, 110
Mono Lake, 53, 99-101, 103, 104, 153
Mörner, Nils-Axel, 98, 152
Morrison Formation, 29
Murray Springs, 147
Museum of the Rockies, 41, 65
Mount Rainier, 96
Mutation, 20, 46, 48, 51, 59, 60, 62, 64, 118, 129, 130

Nagasake, 61, 130
Nanodiamonds, 145-150, 153
Naptha, 124
NASA, 154, 155
Nautilus, 56
Natura non facit saltum, 30
Neanderthal, 73, 103, 165
Neftel, A., 52

188 INDEX

Nellis, W. J., 159
Neptune, 136
Neutron bomb, 116, 129
New Mexico, 149
Newton, Sir Isaac, 90
Nile River, 104
Nitrile, 134
Nitrogen, 49, 54, 85, 114, 115, 144
Northern Lights, 77, 78, 82, 92, 153, 154
North Pole, 51, 76, 90-93, 104
Noyes, Robert, 75
Not by Fire but by Ice, 88, 164
Nuclear, 76, 117, 119, 132
  attack, 59, 61, 75
  accelerator, 83, 84
  collision, 62
  electron capture, 49, 114
  explosion, 79, 106, 107, 113, 114, 116-119, 123, 124, 128, 130, 142, 143
  irradiation, 147
  physicist, 140, 145, 164
  reaction/reactor, 75, 131, 140, 141, 148

Ocean temperatures, 42, 43
Officer, Charles, 113
Oil, 124, 125, 128, 131, 134-136, 139, 143, 157, 159-163
Oil shale, 159
Olsen, Nils, 194
O'Neil, Shaquille, 34
Orbital eccentricity (also orbital stretch) 89
Ordovician, 25, 43
Osmium, 45, 112, 113, 117, 122

Padian, Kevin, 41
Pagel, M., 142
Paw prints, 156
Peking man, 73
Pentaceratops, 67
Permian, 28, 43, 110, 111, 118
Peterman, Zell, 118
Peterson, K. R., 127
Petroleum, 133, 135, 137
Pitch Lake, 125
Pliocene, 110
Pleinsbachian, 43
Pleistocene, 36
Polaris, 91
Polarity reversal (see magnetic reversal)
Potassium, 116
Precambrian, 110, 138, 140
Precession of the equinoxes, 51, 89-91, 95, 153, 167
Prévot, Michel, 95
Pterosaur, 71

Quetzacoatlus, 67

Rad, 61, 62
Radiation, 28, 48, 58-63, 67, 72, 81, 84, 114, 119, 129, 131
Radioactive/radioactivity, 48, 49, 52, 58, 61-64, 74, 84, 88, 89, 96, 102-104, 114-116, 118, 128, 131, 135, 139, 140, 147, 149-151, 154, 156, 161, 163, 166, 167
Radiocarbon, 114-116, 135, 151
Radiogenic, 51, 87, 88, 115, 116
Rainwater basin, 150

Raisbeck, G. M., 53, 97
Rampino, Michael R., 101
Raton Basin, 111
Raup, David, 31, 33, 139
Ray, Dixie Lee, 54
Reversed polarity (see magnetic reversal)
Ries Crater, 142
Rieskessel Crater, 108
Rising reefs, 103
Roperch, P., 99
Royal Terrell Museum, 73, 157
Rudistid, 31, 42
Ryan, W. F., 137

Sagan, Carl, 87, 125, 133, 134
Saint Lawrence River, 104
Salinas, 150
Saltationist, 20
Saturn, 134
Scott, Donald E., 136
Seismosaur, 68
Sepkoski, J. J. Jr., 28
Sharks, 25
Sibeck, David, 154, 155
Siberia, 106, 107
Silica glass, 121, 157
Silurian, 25, 43
Simpson, John F., 48
Sluggards, 57
Smith, Peter J., 92
Smit, Jan, 129
Solar
　flare, 77, 78, 82
　system, 93, 95, 136, 161
　wind, 43, 44, 47, 53, 54, 77, 80, 81, 84, 154-156, 158, 162, 163

Soot, 43, 44, 53, 54, 113, 123, 124, 137, 155, 158, 160, 162, 163
South Pole, 80, 81
Species, 20, 21, 25, 29, 33, 36, 37, 40, 56, 57, 59, 71, 102, 103, 120, 165
Spencer, L. J., 121
Squishy blobs, 24
Stanford Linear Accelerator Center, 83
Stanley, Steven M., 111, 138
Standing trees, 126, 162
Steen River, 142
Steens Mountain, 95
Steiner, J., 88
Stevns Klint, 43
Stoner, Joseph, 92
Strontium, 51, 52, 54, 64, 88, 102, 113, 118, 122, 162, 163, 167
Substorms, 153
Sudbury structure, 142
Sundfold, Scott, 137
Supernova, 53, 112, 153, 154
Suplee, Curt, 76
Sun, 75-80, 83-85, 88-90
Sunspots, 78-80, 84, 85
Survival of the fittest, 36, 56, 72

Tar, 127, 137, 139, 162
Tar pit, 124
Tatoosh intrusion, 96
Tektite, 108-110, 122, 123, 137
Tertiary, 40, 72, 118
Thecodont, 28, 29
THEMIS, 154, 155

Thomas, Rob, 113
Titan, 134-136, 162
Topping, William, 149
Tortoise, 57
Triassic, 28, 29, 43, 118
Triceratops, 66
Trilobite, 24, 28
Tritium, 49, 116
Topping, William, 148, 149
Tschudy, Robert, 41
Tullock Formation, 160
Tunguska, 105-108, 110, 116, 117, 120, 132, 146, 149, 155, 165
Twiddling their evolutionary thumbs, 29
Tyson, Mike, 35

Uffen, Robert, 46-48, 60
Ultraviolet rays, 61, 62
Universidad Formation, 124
Uranium, 59, 88, 131, 132, 139-142, 148, 160, 162, 167
Uranus, 136
USGS, 151, 157, 160

Valet, J. P., 95
Vega, 91
Velikovsky, Immanuel, 135
Vredefort Dome, 142

Waddington, C. J., 48
Watkins, N. D., 48, 99
Ward, Peter D., 56, 72
Warwick-Smith, Simon, 148
Wentworth, Carl M., 51
West, Allen, 145, 146, 148
Whale, 65
Winkler, Gisel, 116
Wolbach, Wendy, 44, 113, 114, 123, 124, 158
Wolfe, Jack, 31
Woods Hole Oceanographic Institute, 33
Wright, Giles, 121

X-ray, 61

Yellowstone, 103
Younger Dryas, 146
Yucatán Peninsula, 109

Zeus, 71
Z coal, 111, 159, 160
Zolotov, A. V., 117

See author's website at
www.iceagenow.com

Sugarhouse Publishing, P.O. Box 435, Bellevue, WA 98009
To order call (425) 821-7372